Green Materials Obtained by Geopolymerization for a Sustainable Future

Petrica VIZUREANU

Dumitru-Doru BURDUHOS-NERGIS

Published by **Materials Research Forum LLC**
Millersville, PA 17551, USA

Published as part of the book series
Materials Research Foundations
Volume 90 (2021)
ISSN 2471-8890 (Print)
ISSN 2471-8904 (Online)

Print ISBN 978-1-64490-112-0
ePDF ISBN 978-1-64490-113-7

Distributed worldwide by

Materials Research Forum LLC
105 Springdale Lane
Millersville, PA 17551
USA
http://www.mrforum.com

Printed in the United States of America
10 9 8 7 6 5 4 3 2 1

Table of Contents

List of Notations and Abbreviations iv

1. Introduction 1

2. State-of-the-Art on Obtaining Geopolymers from Mineral Waste............6
   2.1.   General information 6
   2.2.   Raw materials for geopolymers 9
         2.2.1 Natural minerals 9
         2.2.2 Wastes...10
         2.2.2.1 Power plant ash..10
         2.2.2.2 Red mud...11
         2.2.3 Alkaline activators..13
         2.2.3.1 Sodium hydroxide (NaOH) 16
         2.2.3.2 Sodium silicate 16
   2.3.   Obtaining methods of geopolymers...17
         2.3.1. The influence of the components on the
         geopolymers characteristics...19
         2.3.2. The effect of obtaining process parameters.....................22
   2.4.   Geopolymers characteristics 24
         2.4.1. Effect of acid attack on geopolymers.............................24
         2.4.2. Dimensional stability and durability 27
         2.4.3. Permeability..27
   References 29

3. General Objectives and Methodology of Experimental Research 44
   3.1.   Research objectives...44
   3.2.   Experimental methodology..46
   3.3.   Analyzing methods and equipment's used in the
       experimental plan 47
         3.3.1 Samples and raw materials drying 47
         3.3.2 Particle size separation 48
         3.3.3 Quantitative measurements 50
         3.3.4 Components mixing 52
         3.3.5 Chemical and structural characterization methods 53
         3.3.6 Energy-dispersive X-ray spectroscopy............................54
         3.3.7 X-ray fluorescence spectroscopy 55
         3.3.8 Microstructural analysis by optical microscopy 57
         3.3.9 Microstructural analysis by scanning electron
         microscopy 60

3.3.10 Mineralogical analysis by means of X-Ray diffraction 62
3.3.11 Mineralogical characterization by Fourier transform infrared spectroscopy 64
3.3.12 Pore size relative distribution 65
3.3.13 Methods used for physical-mechanical evaluation 69
3.3.13.1 Compressive strength evaluation....................69
3.3.13.2 Flexural strength evaluation 72
3.3.14 Evaluation of thermal behavior................................73
References 75

4. Design and Development of Geopolymers Based on Thermal Power Plant Ash...**78**
4.1.   Raw materials characterization 78
    4.1.1. Thermal power plant ash characterization 78
    4.1.2. Glass powder characterization.......................79
    4.1.3. Characterization of the natural aggregate...........80
4.2.   The geopolymers designing 82
4.3.   Obtaining of coal-ash based geopolymers 85
    4.3.1. Solid component preparing 85
    4.3.2. Liquid component preparation 85
    4.3.2.1. Sodium hydroxide solution (NaOH) 86
    4.3.2.2. Sodium silicate solution (Na_2SiO_3) 86
    4.3.3. Components mixing stage 86
4.4.   The process flow of coal ash based geopolymers obtaining 87
References 90

5. Chemical and Structural Analysis of Geopolymers **95**
5.1.   Chemical analysis using EDAX.............................95
5.2.   Structural analysis using optical and scanning electron microscopy..97
5.3.   Mineralogical analysis by X-ray diffraction.............108
5.4.   Structural characterization by FTIR spectroscopy 117
References 123

6. Physical-Mechanical Analysis of Geopolymers....................**126**
6.1.   Setting time 126
6.2.   Relative pore size distribution 128
6.3.   Compressive strength...................................137
6.4.   Flexural strength 138
6.5.   Thermal behavior evaluation of geopolymers 139

References 154

7. Sustainability with Geopolymers...158
    7.1.   Geopolymers technology for green cities 158
    7.2.   Environmental impact of geopolymers......................................159
    7.3.   Applications 162
           7.3.1. Civil engineering 162
           7.3.2. Geopolymers as multifunctional materials......................163
    7.4.   Short overview on the geopolymers engineering applications...165
    References 172

8. Green Materials Tendencies for a 176
    8.1.   Tendencies in geopolymers technology......................................176
           8.1.1. Geopolymers advantages......................................179
           8.1.2. Geopolymers disadvantages 179
    8.2.   Tendencies in ferrock technology...181
    8.3.   Tendencies in Sorel cement technology 183
    8.4.   Tendencies in cork technology 186
    8.5.   Tendencies in sugarcane bagasse technology..............................188
    8.6.   Tendencies in oriented strand board technology 190
    References 191

9. Conservation Potential by Aluminosilicates Recycling............................194
    9.1.   Recycling potential and raw materials conservation 194
    9.2.   Replacing potential by mine tailings use 197
    9.3.   Replacing potential by red mud use...200
    9.4.   Replacing potential by fly-ash use..201
    References 203

About the Authors...205

List of Notations and Abbreviations

(Si-O-Al) – *sialate molecule (chemical compound of silicon, aluminium and oxygen)*;

(-Si-O-Al-O-) – *polysialate molecule;*

(-Si-O-Al-O-Si-O-) – *polysialate-siloxo molecule*;

(-Si-O-Al-O-Si-O-Si-O-) – *polysialate-disiloxo molecule*;

100FA – geopolymer sample with coal ash in 100 wt., % of solid component;

70FA_30PG – geopolymer sample with coal ash in 70 wt., % and glass powder in 30 wt., % of solid component;

30FA_70S – geopolymer sample with coal ash in 30 wt., % and sand particle in 70 wt., % of solid component;

15FA_15PG_70S – geopolymer sample with coal ash in 15 wt., %, glass powder in 15 wt., % and sand particle in 70 wt., % of solid component;

BA – *bottom ash*;

C-A-S-H – *Calcium Alumino Silicate Hydrate*;

CPMG – Carr–Purcell Meiboom–Gill analysis technique;

C-S-H - *Calcium Silicate Hydrate*;

DTA – *Differential Thermal Analysis*;

EDS –*Energy-Dispersive X-ray Spectroscopy*;

FA – *fly ash*;

FTIR – *Fourier-Transform Infrared Spectroscopy*;

MO – *optical microscopy*;

N-A-S-H –*Sodium Aluminosilicate Hydrate*;

NR – *red mud, waste resulted from alumina extraction from bauxite by means of Bayer process*;

PG – *glass powder, waste resulted from glass tanks crushing;*

RMN – *Nuclear Magnetic Resonance*;

S – sand, river sand particles;

SEM – *Scanning Electron Microscope*;

TGA – *Thermogravimetric Analysis*;

TG-DTA – *Simultaneous Thermogravimetric Analysis*;

XRD – *X-ray diffraction*;

XRF – *X-ray fluorescence*;

ZG – *granular furnace slag*;

OSB – oriented strand board;

1. Introduction

The large volume of waste generated nowadays represents a serious issue for both the environment and the population living near the storage areas (dumps) of these by-products resulting from technological processes [1–4]. Another problem associated with construction materials obtaining, mainly Ordinary Portland Cement (OPC) based concrete, is related with the exploitation of non-renewable raw materials, limestone and clay, which also involve high energy consumption for transportation and processing [5,6]. The overall consumption of non-renewable resources is around 20,000 million tons/year and an annual increase of 4.7% is expected [7]. More than a third of this consumption is related to concrete production, being the most used construction material on earth, currently exceeding 10 km^3/year [8]. As the cost of raw materials depends very much on transport distances, this leads to their exploitation in areas as close as possible to construction sites, thus producing an accelerated increase in the number of quarries and increasing the impact on biodiversity. Therefore, in order to reduce the consumption of raw materials and economic resources, in recent years there has been a global emphasis on the reuse of materials [9–13]. A simple technique of high interest, which can convert aluminosilicate wastes in construction materials with properties similar to OPC concrete, is geopolymerization [14–17]. This is the chemical reaction of dissolving a material rich in silicon and aluminum oxides under the action of an alkaline solution and forming a tetragonal structure of Si-O-Al following the partial removal of water. Therefore, a geopolymer is an oxide material based on aluminum and silicon groups, chemically balanced by Na^+, K^+ etc. atoms, which is formed following the geopolymerization reaction [18,19].

Due to the simplicity of the obtaining process, their final properties, but also the abundance of raw materials, since 1979 [20] this oxide materials have participated in cutting-edge research and applications in the developing of thermal insulation, fire-resistant materials, construction materials, industry metallurgy, decorative objects, refractory linings, repair and consolidation of infrastructures, automotive industry, aeronautical industry, cements and concretes, encapsulation of radioactive and toxic waste etc. [21-25]. However, the main field of use of geopolymers is the construction industry as environmentally friendly concrete, with low energy consumption and low CO_2 footprint compared to conventional Portland cement concrete [26]. Therefore, geopolymerization is an advantageous technique for obtaining environmentally friendly materials that have comparable or superior properties to those of conventional materials but uses mineral waste as a source of raw material.

The publication of scientific papers in the field of geopolymers began in 1988 with the work of Davidovits [19], their number continued to be small until 2004-2006, due to the few researchers interested in these materials. However, after 2007 the number of publications increased by hundreds, exceeding 250 in 2013, and in 2019 the number of publications was over 850. Currently, geopolymers are studied in many laboratories around the world, and the number total number of publications in this field exceeds 4500 works. According to this, any material rich in aluminum and silicon, which can be dissolved under alkaline condition, can be used to obtain a geopolymer. Globally, several wastes have been identified with potential for geopolymerisation, such as fly ash, red mud, blast furnace slag etc. After the chemical reaction between the solid raw material (wastes) and the alkaline activator (a solution of sodium silicate or potassium silicate and sodium hydroxide), inorganic material with a structure similar to that of zeolite is obtained.

It is well known that the ordinary Portland cement-based materials present multiple disadvantages related to the manufacturing technology, mechanical/chemical properties or the environmental impact. During the last decade, the research efforts in the field of building materials have been focused on obtaining environmental-friendly materials (eco-friendly). Depending on the area where the final material will be used, the geopolymers are designed with tailored properties. For example, the porosity and the integrity of geopolymers structure depends of the Si/Al ratio from the system, while the heavy metals absorption capacity depends mostly on the reactive functional groups' concentration, therefore when high porosity and high content of functional groups are required, those two parameters must be considered. Usually, in the case of geopolymers, the most reactive functional groups are hydroxyl groups, according to literature, those groups are capable of taking part in many chemical reactions as well as ion exchange processes. Up to now, the functional geopolymer materials, especially kaolin-based geopolymers, have been successfully used in multiple applications. However, there are still many challenges to solve and much more effort is needed to further improve the quality and function of kaolinite-based materials, while in case of fly-ash geopolymers, currently no high dimension samples have been developed as functional materials. Therefore, up to date, the geopolymers used as functional materials are connected with natural raw materials (kaolin) consumption or inefficient structures (in case of fly ash-based geopolymers). Moreover, comparing with the building materials for the civil construction industry, the geopolymer composites are easy-to-prepare and can be synthesized at room temperature (low energy consumption) using as raw materials industrial waste, i.e. without energy consumption for its curing process or raw materials processing, while for bricks or different types of clay products the burning stage takes place at elevated temperature.

The book contains a complex and interdisciplinary study in the field of physics, chemistry, materials science and civil engineering on oxide materials based on mineral wastes, known as geopolymers. At the same time, it contains advanced notions about the materials design, development and characterization of unique materials based on the geopolymerization process. Their main advantage is the use of indigenous mineral waste as a source of raw material, as well as its physical and mechanical characteristics comparable to those of classical Portland cement-based materials.

The book is relevant for fundamental and applied research in the field of materials engineering because it shows the obtaining of ecofriendly geopolymers with industrial applications. It also describes how to design and develop different types of geopolymers that use as reference the availability of indigenous mineral wastes and the literature in the field of geopolymers. Moreover, these geopolymers are obtained according to original process flow and characterized, by specific analyzes in laboratory conditions, from a chemical, structural, physical-mechanical and thermal point of view. The geopolymers analysed use thermal power plant ash for matrix synthesis and two types of reinforcing elements. The main difference between the studied geopolymers is given by the type and proportion of reinforcing elements. In order to obtain a material of industrial interest, a simple technological process has been designed that leads to the formation of a geopolymer based on mineral waste with optimal mechanical properties.

The first chapter entitled *"Introduction"* presents introductory concepts, about geopolymers as green materials with a low carbon footprint.

The second chapter, *"State-of-the-Art on Obtaining Geopolymers from Mineral Waste"*, presents fundamental concepts related to materials based on mineral waste or natural raw materials that are obtained by geopolymerization. Also, it contains general information about the methods of obtaining geopolymers and their applications in industry.

The third chapter, *"General Objectives and Methodology of Experimental Research"*, presents the research focus and the characterization methods of the obtained geopolymers from (if) chemical point of view by X-ray fluorescence spectroscopy and X-ray spectroscopy by energy dispersion, (ii) structural point of view by optical microscopy, scanning electron microscopy, nuclear magnetic resonance and X-ray diffraction, (iii) physical-mechanical point of view by compressive strength, bending strength and Vicat tests and (iv) thermal point of view by thermogravimetric analysis and differential thermal analysis. Moreover, this chapter briefly describes the techniques and equipment used to analyze the obtained materials.

The fourth chapter, "*Design and Development of Geopolymers Based on Thermal Power Plant Ash*", presents the design and development of original compositions of geopolymers made by recovering indigenous mineral wastes, as well as a detailed description of the technological flow of production, which takes into account most of the parameters that affects the chemical reaction of geopolymerisation.

In order to obtain the geopolymer materials, a production method was designed which includes the following steps: preparation of the liquid and solid component, mixing of the components, obtaining the geopolymer binder, pouring the binder into shapes and drying/obtaining the geopolymers.

The fifth chapter, "*Chemical and Structural Characterization of Geopolymers*", describes the chemical aspects on which the obtained geopolymers are based, also, it contains the experimental results obtained during chemical, microstructural and mineralogical analysis.

The sixth chapter, "*Characterization of the Physical and Mechanical Properties of Geopolymers*", presents the physical-mechanical properties and thermal behavior of materials obtained through standardized methods used for conventional oxide materials.

The seventh chapter entitled "*Sustainability with Geopolymers*" presents successfully application of geopolymers as a replacement for conventional materials and indicates possible future directions for these revolutionary materials which will participate in cutting-edge research and applications due to their tailored properties. Also, it includes a short overview on the engineered properties of coal-ash based geopolymers.

The eighth chapter, "*Green Materials Tendencies for a Sustainable Future*" describes the sustainability concept considering the industrial development introduced by the use of green materials. Green materials are considered those products that use as raw materials regenerable sources and emphasis on reducing the use of hazardous substances in the design, manufacture and application. In this chapter are presented the base concepts and the research direction in the field of different types of ecofriendly materials, such as geopolymers, ferrock, cork, sorel cement, bagasse products and oriented strand board.

The last chapter entitled "*Conservation Potential by Aluminosilicates Recycling*" contains information regarding the recovery and reuse of recyclable resources, aiming to resolve the contradiction between the requirements of the economic growth process and the restrictive nature of resources. Therefore, the availability of different types of aluminosilicate sources are presented.

Following the design, elaboration, characterization and evaluation of the thermal behavior of the materials described in the present book, it was concluded that geopolymers are materials composed of several phases (sodalite, quartz, corundum, anorthite, hematite, calcium hydroxide, calcium carbonate, etc.) with different molecular groups and porous structure (Si-O, Si-O-Al, $\equiv$Si-OH, etc.). From the mechanical properties point of view, the geopolymers obtained show compressive strength values corresponding to the C8/10 class of concrete, which is mainly used for foundations, or to the C16/20 class of concrete that can be used for the obtaining of resistance structures (pillars, beams, belts or slabs) and foundations. In terms of compressive strength, the geopolymers from class C8/10 are ideal materials for ground floor houses or buildings with one level. From the thermal behavior point of view, the presented geopolymers are stable in the temperature range (25÷340) °C and become unstable above this temperature due to the decomposition of some hydroxides (calcium hydroxide, goethite and aluminum hydroxide etc.).

2. State-of-the-Art on Obtaining Geopolymers from Mineral Waste

2.1. General information

In the early 1950s, the Ukrainian researcher Viktor Glukhovsky developed and analyzed new materials for cement known as "silicate soil concretes" and "soil cement". Later, in 1979, when Joseph Davidovits introduced the concept of geopolymers, the terminology and definition of these materials became much more diverse and often conflicting.

Geopolymers are inorganic materials based on silicon and aluminum oxides, chemically balanced by group I alkaline ions [1]. These are rigid gels, created relatively, under normal conditions of temperature and pressure, which can then be transformed into crystalline or ceramic materials, similar to zeolites [2]. A geopolymer is a very long reticular chain with groups of silicon $(SiO_4)^{4-}$ and specific tetragonal network of aluminum oxide $(AlO_4)^{5-}$ resulting from the exothermic process of some oligomers. The bonds between these tetrahedral structures are balanced by alkaline ions of K^+, Na^+ or Li^+ [3,4].

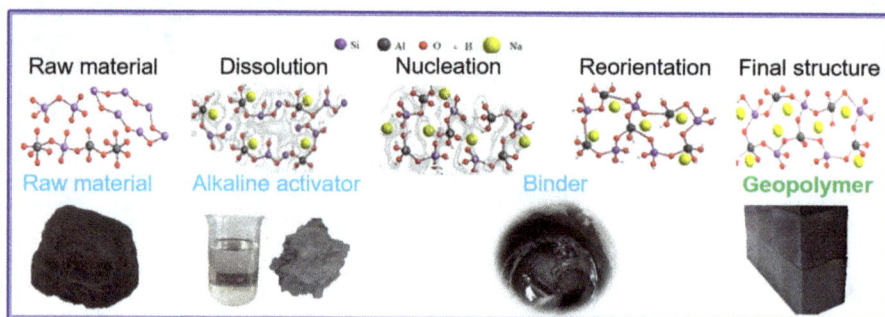

Figure 2.1. Geopolymerisation stages [5–8].

The geopolymers are formed by a chemical reaction, known as geopolymerization, which occurs as a result of mixing at least two constituents (a solid rich in silicon and aluminum oxides and an alkaline solution). Geopolymerization (Figure 2.1) begins with the dissolution of the silicon, aluminum and calcium (if any) hydrates from the solid material (raw material) under the action of the alkaline activator. In the second stage, nucleation, oligomerization, polymerization and polycondensation occur, therefore, the groups of atoms are reoriented and groups called polysialates are created. The term polysialate was introduced by J. Davidovits being a generic name of oxide groups of alumino-silicates (Table 2.1) [9]. In other words, after the dissociation of the Si-O-Si and Si-O-Al compounds, highly reactive Al^{3+} and Si^{4+} species are released which, following the interaction between them and oxygen, form oligomers of $(SiO_4)^{4-}$ and $(AlO_4)^{5-}$ which will further create 3D polymer chains of Si-O-Al-O, along with water elimination [9,10]. The type and structure of the compounds from the polymer chains created are influenced by the Si to Al ratio, based on the empirical relationship (2.1) [9]:

$$R^+{}_v \cdot \{- (SiO_2)_x\text{-}AlO_2\text{-}\}_v \cdot aH_2O \qquad\qquad (2.1)$$

where: R^+ - the alkaline cation from the activator solution (Na^+, K^+ etc.); v – polymerization degree; x – Si to Al ratio, a – the number of water molecules (water quantity).

The ratio (x) between Si and Al can have values in the range of 1 to 300 if the value of (x) is less than 3 ($x < 3$) then the geopolymer will possess high adhesion and flexibility properties due to the linear 2D polymeric structure. As the ratio (x) increases, the fragility of the final structure increases, and its reticular network becomes 3D (Figure 2.2) [1,5,11]. Along with, it was observed that with the increase of the silicon oxide to aluminium oxide ratio, an increase in the setting time will appear [11,12].

Table 2.1. Types of polysialates existing in geopolymers [6,13,14].

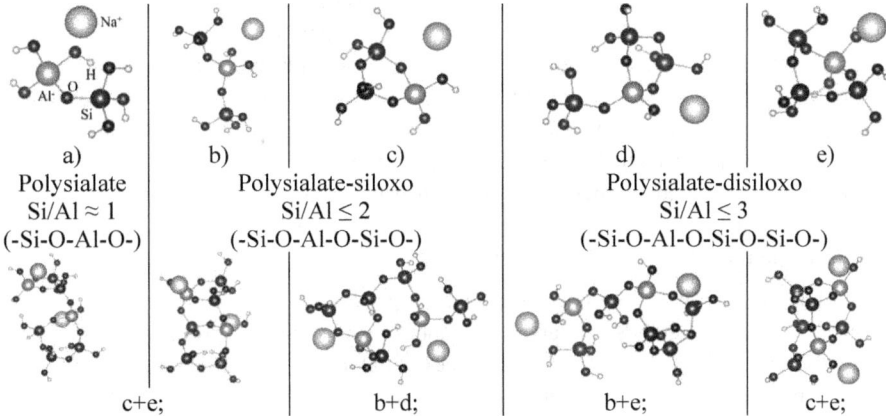

a)	b)	c)	d)	e)
Polysialate	Polysialate-siloxo		Polysialate-disiloxo	
Si/Al ≈ 1	Si/Al ≤ 2		Si/Al ≤ 3	
(-Si-O-Al-O-)	(-Si-O-Al-O-Si-O-)		(-Si-O-Al-O-Si-O-Si-O-)	
c+e;	b+d;		b+e;	c+e;

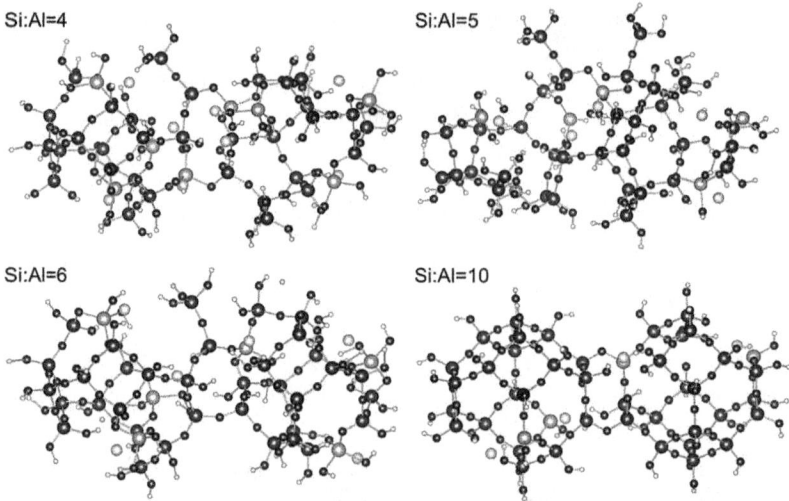

Figure 2.2. 3D structure of the geopolymers depending on the Si to Al ratio [15].

Due to its physical, chemical and mechanical properties, geopolymers have participated in cutting-edge research and became of great interest for many areas of industry. In the beginning, these were developed as high fire resistance materials ideal for civil engineering, but later they were introduced in the automotive, ceramic, metallurgical, aerospace etc. industries, as substitutes for conventional oxide materials or polymers. The fast-developing and increased interest in these materials in the industry were related to their special characteristics, such as the self-healing mechanism [16,17]. This mechanism consists of the fact that some defects (cracks) are repaired (disappear) as a result of the continuation of the geopolymerization reaction between the unreacted particles and the gel remaining in the pores. Therefore, the gradual increase of mechanical properties over time is due to the self-healing special feature.

Even though geopolymers possess superior properties and, at the same time, can be obtained by simple methods, the largest amount of them is obtained from natural minerals (kaolin, clay etc.) [13,18,19]. Therefore, it is essential to design and develop geopolymers that use mineral waste, especially indigenous waste, as a source of raw material.

2.2. Raw materials for geopolymers

Any geopolymer can be divided into two main constituents, the base material and the activator (an alkaline liquid). The main constituent is the base material, this must be rich in silicon and aluminum oxides and can be a natural mineral such as clay, kaolin etc. or waste, such as thermal power plant ash, red mud, slags etc. [20–27].

2.2.1 Natural minerals

The first natural mineral used to obtain geopolymers was kaoline. In 1972, the ceramists team of M. Davidovits and J.P. Latapie successfully create water-resistant ceramic bricks at temperatures below 450 °C, i.e. without burning [6,9]. The experiment was besd on the fact that, at 150 °C the kaoline, a clay component, reacts with sodium hydroxide and polycondenses into hydrated sodalite (a tecto-aluminum-silicate) or hydro-sodalite [1].

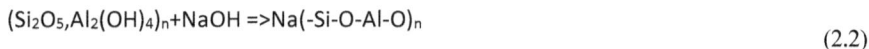

$$(Si_2O_5, Al_2(OH)_4)_n + NaOH => Na(-Si-O-Al-O)_n$$

(2.2)

Kaoline hydro-sodalite

In 1969, Besson, Caillèreand and Hénin from the Natural History Museum in Paris succeeded in synthesizing hydrosodalite (ec. (2.2)) from several types of phyllosilicates (kaolin, halloysite) in concentrated NaOH solutions [6].

Kaolin ($Al_2O_3 \cdot 2SiO_2 \cdot 2H_2O$), also known as kaolinite, is a clay based on oxides of Si, Al, Fe and Ca (Table 2.2), with an earth-like texture that can be easily molded or sculpted [28]. This clay is a layered silicate mineral, with a tetrahedral layer of silicon oxide $(SiO_4)^{4+}$ bonded by oxygen atoms to an octahedral layer of aluminum oxide (AlO_6).

Table 2.2. Typical chemical composition of kaolin [23].

Oxide	SiO_2	Al_2O_3	Fe_2O_3	CaO	Loi.*
[%], wt.	54.47	43.95	0.52	0.34	0.72

*Loi. – loss on ignition;

In order to become suitable as a source of raw material for geopolymers, the kaoline is converted, by calcination, into metakaolin. In normal environmental conditions, kaolin is stable, however, when it is exposed to temperatures between (650 ÷ 900) °C, it loses up to 14% of its mass by breaking hydroxyl bonds and losing water [13]. This heat treatment destroys the structure of kaolin (Figure 2.3), so the layers of aluminum and silicon oxide lose their order resulting in a highly reactive material, usually in form of powder, known as metakaolin.

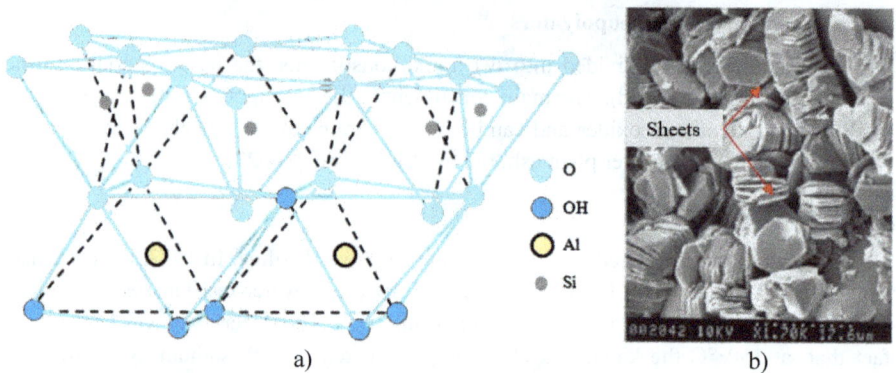

O
OH
Al
Si

a)   b)

Figure 2.3. Kaolin. a) chemical structure; b) microstructure [29,30].

Metakaolin is a highly reactive transition phase, with a high content of silicon and aluminum, which can be used as a base material for the manufacture of mortars due to its pozzolanic capacity [31].

2.2.2   Wastes

According to the literature, any compound rich in silicon and aluminum that can be dissolved in an alkaline solution can be used as a raw material to obtain geopolymers. Therefore, their manufacture was carried out by mixing several categories of mineral waste, such as power plant ash, red mud, blast furnace slag, rice straw ash, wheat straw ash etc. The most commonly used of these are power plant ash and red mud, due to the spread, as well as the fact that there are huge quantities stored worldwide [32–35].

2.2.2.1 Power plant ash

Thermal power plant ash is a powdery product that results from the process of decomposition by oxidation (burning) of coal introduced into the burning chamber of a thermal power plant. Due to the heat released by its combustion, the temperature in the burning chamber reaches about 1700 °C [36,37]. At this temperature, non-combustible inorganic minerals (quartz, calcite, gypsum, goethite, clay minerals etc.) melt and, by atomization, take the form of droplets [38,39]. Depending on their size, some of the drops are evacuated at the top of the furnace, in suspension in the flue gases, and the others (the big drops) falls at the bottom of the furnace [40]. The droplets thus evacuated become spherical glass-like particles known as fly ash (Figure 2.4). This powder is being collected from the flue gases by means of mechanical and electrostatic precipitators (filters). The drops that reach the bottom of the furnace have a porous structure and irregular shapes being known as bottom ash (Figure 2.5) [41]. Due to the high content of silicon and aluminum oxides from their composition, both fly ash and bottom ash are suitable as raw materials for geopolymers [42,43].

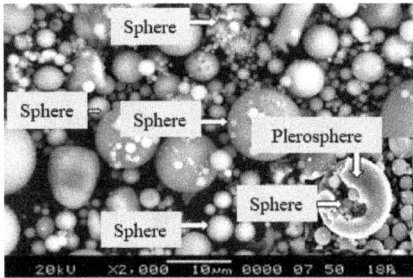

Figure 2.4. Fly ash morphology [44].   *Figure 2.5. Bottom ash morphology* [45].

The performance of power plant ash in geopolymers is strongly influenced by its chemical compounds, phases and constituents [46,47]. The chemical and mineralogical composition depends to basically on the composition of the coal, taking into account the fact that there are many types of coal (anthracite, bituminous lignite or sub-bituminous lignite) [48,49]. According to the specifications of the standard for calcined or uncalcined fly ash and natural pozzolanic materials (ASTM C618), the ash discharged at the top of the furnace is divided into two classes. Class F which includes fly ash produced by burning bituminous coal or anthracite that meets the chemical condition (ec. (2.3)).

$$SiO_2 + Al_2O_3 + Fe_2O_3 \geq 70\ \% \qquad\qquad\qquad\qquad (2.3)$$

$$SiO_2 + Al_2O_3 + Fe_2O_3 \geq 50\ \% \qquad\qquad\qquad\qquad (2.4)$$

And Class C which includes fly ash produced by burning sub-bituminous coal or lignite that meets the chemical condition (ec. (2.4)). Some types of class C ashes may contain more than 10% calcium oxide [50].

2.2.2.2 Red mud

Red mud is the main waste resulting from the refining of bauxite, in order to obtain alumina, through the Bayer process (Figure 2.6) [51]. A bauxite refinery normally produces twice as much red mud as alumina [52]. This ratio depends on the type of bauxite used and the extraction conditions. This waste consists mainly of metal oxides of Al, Fe, Ti etc. The red color comes from iron oxides which represent about 60 % of the mass of the mixture. Due to a large amount of NaOH used to dissociate the compounds, red mud has a basic pH between 10 and 13 [53].

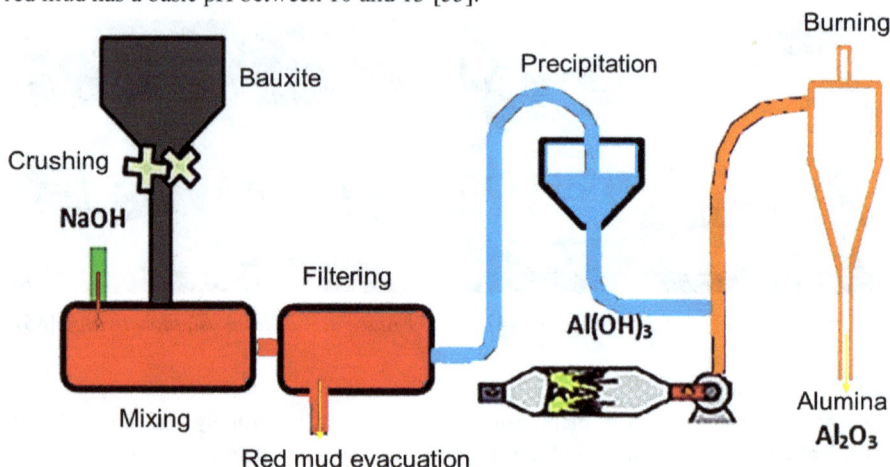

Figure 2.6. Schematic representation of the Bayer process [54].

There are over 60 refineries [52,53] worldwide that use the Bayer process to produce alumina from bauxite ore. In order to extract alumina through this process, the soluble part of the bauxite ore is dissolved with sodium hydroxide under conditions of high temperature and pressure. After removing the insoluble residues (red sludge), a solution of sodium aluminate results in which, as the temperature decreases, the aluminum hydroxide precipitates in the solution. After cooling, part of it is kept for the precipitation of the next batch, and the rest is passed to the final stage. The last stage of the process, calcination, consists of heating the batch to over 1000 °C in rotary kilns in order to obtain aluminum oxide (alumina). The alumina content of normally used bauxite exceeds 50 % and is found in the form of complex chemical compounds such as gibbsite ($Al(OH)_3$), boehmite ($AlO(OH)$) or diasporite ($AlO(OH)$). The residues inevitably have a high concentration of iron oxide (Table 2.3) which gives to this waste a characteristic red color. Additionally, a small amount of sodium hydroxide doesn't dissolute during the process and end by being eliminated with insoluble residues, causing its high pH [53].

Table 2.3. Chemical composition of red mud [55].

Oxide	Fe_2O_3	Al_2O_3	TiO_2	CaO	SiO	Na_2O
[%], wt.	5-60	5-30	0.3-15	2-14	3-50	1-10

The storage methods of red mud have been changed substantially since the construction of the original refineries. In the early ages, it was practiced to pump tailings sludge (insoluble residues and water), at a concentration of about 20 % solid, in lagoons or ponds sometimes created in former bauxite mines or impoverished quarries [51]. In other cases, earth dams were built or various ponds were used nearby, but the breaking of dams or the banks of ponds resulted in catastrophic consequences, both for the environment and for the surrounding population [56,57].

2.2.3   Alkaline activators

The chemical activation of the selected raw material is an important factor in the production of a material by geopolymerization because the activator is "responsible" for the precipitation and crystallization of aluminum and silicon species present in the binder. OH^- groups act as a catalyst for reactivity, while the metal cations serve to create a stable structure and balance the negative tetrahedral aluminum lattice [5,58]. The initial mechanism of the reaction is driven by the capacity of the solution to dissolve the raw material and to release reactive silicon and aluminum species into the solution [59].

When power plant ash or another material rich in aluminum and silicon is mixed with an alkaline solution, the glassy component is quickly dissolved. However, there is not enough time and space for the formed gel to pass into a crystalline structure. Therefore, the resulting material has a semi-crystalline or semi-amorphous structure [60]. The alkaline solutions used as activators in geopolymers technology can be classified as follows, (R) represents the alkaline cation [5,61]:

- Alcaline, ROH;

- Low acid salts, R_2CO_3, R_2SiO_3, R_3PO_4;

- Silicates, $R_2O \cdot nSiO_3$;

- Aluminates, $R_2O \cdot nAlO_3$;

- Aluminosilicates, $R_2O \cdot nAl_2SO_3 \cdot (2-6)SiO_2$.

Although various studies [62–65] have been performed on the activation of geopolymers with solutions of NaOH, Na_2SiOi, Na_2CO_3, K_2CO_3, KOH or K_2SiO_3, the most commonly used alkaline activators are a mixture of sodium hydroxide (NaOH) or potassium hydroxide (KOH) and sodium silicate (Na_2SiO_3) or potassium silicate (K_2SiO_3) [66–68]. Increasing the concentration of NaOH or KOH in the activation solution improves the dissolution capacity and directly influences the properties of the resulting geopolymer. According to the study conducted by C. Panagiotopulou et al. [69], materials rich in silicon and aluminum dissolve more easily in NaOH solution than in KOH, but geopolymers obtained with KOH solution acquire a higher compressive strength. According to H. Xu et al. [70], Na^+ ions more easily associate with the Si anion and form small oligomers. However, the pairing of large K^+ ions with Si anions produces larger oligomers that positively influence the compressive strength, increases up to 40%, of the geopolymer.

However, the dissolution capacity also depends on the type of dissolved material, according to C. Panagiotopoulou et al. [69] this is maximum for metakaolin and minimum for kaolin. This study highlighted the solubilization capacity of Si and Al atoms, using NaOH or KOH solution, from the main types of raw material for geopolymers. The ordering of the materials according to the mass percentage dissolved by those two solutions is metacaolin> zeolite> slag> thermal power plant ash> puzzolan> kaolin (Figure 2.7).

From the morphological/microstructural point of view (Figure 2.8), a geopolymer activated with a sodium-based solution has a globular-textured matrix and a small number of unreacted metakaolin particles (layered structure) [72]. However, geopolymers activated with a potassium solution have a finer texture and a denser structure [73]. Moreover, according to other studies [74,75] Na^+ based geopolymers possess superior compressive strength, contrary to the expectations based on the microstructural analysis. This may be due to the high activity of Na^+ which leads to a higher dissolution capacity, as well as a better thermal stability of the monomers formed [72].

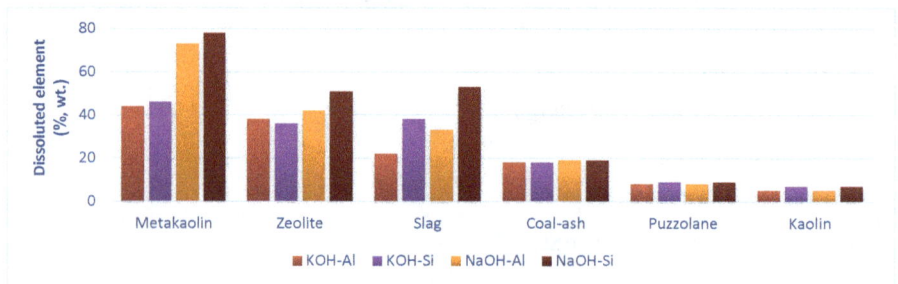

Figure 2.7. The weight percentage of Al and Si dissolution by NaOH or KOH solutions [71].

Although the sodium based alkaline activator has a higher dissolution rate and capacity, it prevents the formation of a homogeneous structure, thus resulting in a porous structure [76,77]. Therefore, the alkaline solution based on sodium is mainly used for economic reasons.

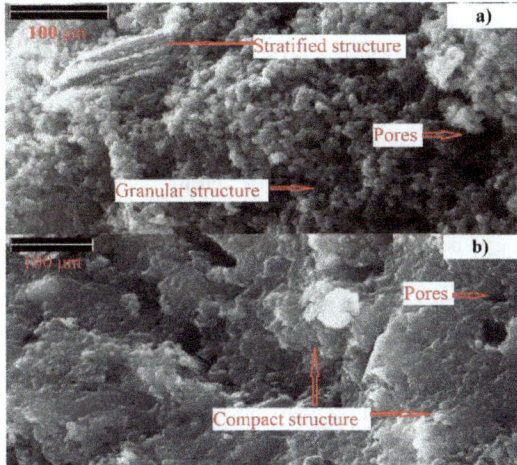

Figure 2.8. The morphology of a geopolymer activated with: a) NaOH solution; b) KOH solution [74,75].

In order to eliminate the disadvantages of each activator and to cumulate their specific advantages, several researchers [78,79] used combinations of activation solutions, such as K_2SiO_3 and NaOH, Na_2SiO_3 and NaOH, K_2SiO_3 and KOH etc. In the D.L.Y. Kong et al. [80,81] was found that geopolymers activated with Na_2SiO_3 condense (harden) faster than those activated with K_2SiO_3 and that those activated with an alkaline solution with metal ions of the same type have superior properties to those activated with mixtures of hydroxides or silicates of sodium and potassium.

In addition to alkaline activators based on Na^+ and K^+, in the study of V. F. F. Barbosa et al. [82], a lithium-based solution was used to obtain halogen geopolymers. Surprisingly, the final material did not show typical amorphous characteristics but mainly showed lithium zeolites, due to the catalytic effect on gel formation and phase separation.

2.2.3.1 Sodium hydroxide (NaOH)

NaOH is an alkaline activator often used for the production of geopolymers because sodium cations are smaller than those of K^+ and can migrate more easily through the network, producing a higher zeolitization [83,84]. In the case of sodium hydroxide, the most important parameter is the molar concentration of the solution, because a concentration too low does not have the capacity to dissolve the solid material, while a concentration too high decreases the capacity of CH (calcium-hydroxide) bonds formed during the gel phase. Besides, the use of an activator with high NaOH concentration leads to the formation of a geopolymer with high mechanical properties at early stages, but over time the excess OH^- groups cause an uneven morphology and negatively influence the durability of the material [84]. However, NaOH activated geopolymers possess high crystallinity resulting in better stability in acidic or sulfated media [85–87].

2.2.3.2 Sodium silicate

Sodium silicate is a generic name for the chemical compounds with the formula $Na_{2x}Si_yO_{2y+x}$ or $(Na_2O)_x \bullet (SiO_2)_y$, such as sodium metasilicate (Na_2SiO_3), sodium orthosilicate (Na_4SiO_4) and sodium pyrosilicate (Na_6Si_2). This compound can be obtained by a hydrothermal process or by mixing and melting.

The hydrothermal process, also known as the wet process, consists in mixing, in several stages (Figure 2.9), of a quantity of silicon dioxide (sand) with a quantity of sodium hydroxide and water. Sodium hydroxide and sand dissolve during the mixing steps, resulting in an unstable white solution. In order to stabilize it, the mixture is passed through a bleaching step (maintenance at high temperature). The bleaching stage results in a highly transparent liquid that must be bottled immediately to avoid contact with oxygen.

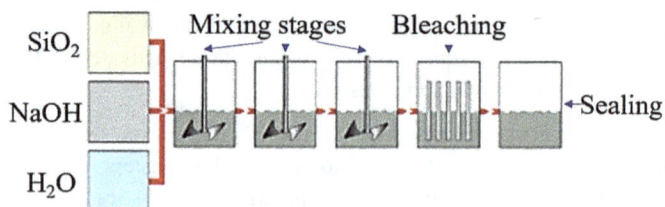

Figure 2.9. Schematic representation of the hydrothermal process for obtaining sodium silicate [88].

The second method of sodium silicate production (Figure 2.10) consists in mixing an amount of silicon dioxide with an amount of sodium carbonate and melting them at a temperature between 1200 °C and 1300 °C. After that, the melt is cooled and mixed with a quantity of water, and then it is heated with steam, mixed, bleached and bottled.

Sodium silicate is rarely used as an independent activator because it does not have sufficient activation potential to initiate the geopolymerization reaction. Therefore, it is usually mixed with a solution of sodium hydroxide, which increases the alkalinity of the combination [89]. The most important parameter of the sodium silicate solution is the mass ratio of silicon dioxide (SiO_2) to sodium oxide (Na_2O) in the solution. Commercially, sodium silicate is available at the ratio (Si_2O/Na_2O) between 1.5 to 3.2 [68,90].

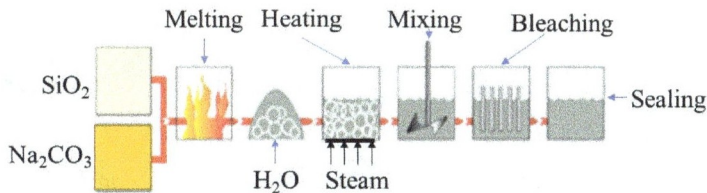

Figure 2.10. Schematic representation of the process of obtaining sodium silicate by the mixing and melting method [91].

In geopolymers, sodium silicate decreases alkaline saturation and increases the strength of the bonds formed between the particles in the composition. Therefore, with the introduction of a large amount of sodium silicate into the activation solution, a marked increase in mechanical properties, in particular compressive strength, is obtained. More than that, this solution increases the resistance of the interface that is formed between the geopolymer matrix and the reinforcing elements [92].

2.3. Obtaining methods of geopolymers

The process of obtaining (Figure 2.11) a geopolymer begins by mixing at least two constituents, the base material and the alkaline activator. The decision to choose the base material is influenced by several factors, such as its cost or availability as well as the scope/application of the resulting geopolymer. In most of the studies, the alkaline activator used is a solution that combines the dissolution capacity of sodium hydroxide and the aggregation capacity of sodium silicate. After mixing the components, the process is followed by a period of hardening under normal environmental conditions ($\approx$ 20 °C) or at slightly high temperatures (<100 °C). During the curing stage, the chemical geopolymerization reaction occurs, which can be divided into three main stages. In the first stage, the solid component is dissolved due to the presence of the activation solution. After removing a small amount of water, the reorientation begins, where the groups of atoms occupy their position in the structure and, at the same time, outlines the solid structure of the geopolymer. Further, the water is almost completely removed and the material is converted to the final shape.

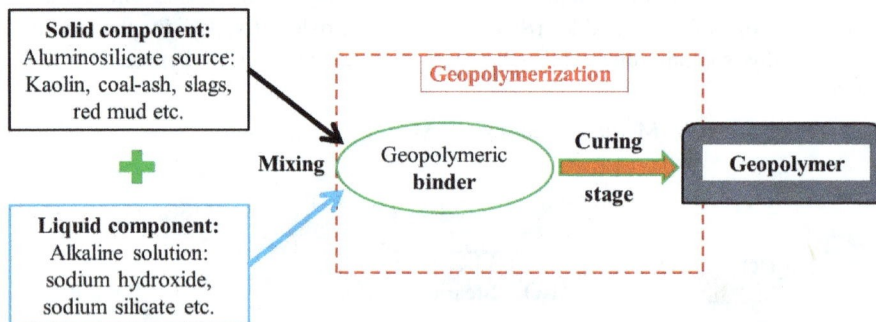

Figure 2.11. Schematic representation of the obtaining process of a geopolymer [92].

Because the geopolymerization reaction is governed by time, after the end of the curing stage it continues at the micro/nanoscopic level due to the reaction between the unreacted particles and the solution in the gel-like pores. This phenomenon gives to the geopolymers the capacity to repair (self-healing) some of the cracks formed by dehydration [11,76].

Both the properties and the quality of geopolymers are strongly dependent on a multitude of factors specific to the mixed components, but also on the parameters of the production process [93–95]. The most important factors, specific to mixed components, that affect the properties of geopolymers are SiO_2/Al_2O_3 ratio, R_2O/Al_2O_3 ratio, SiO_2/R_2O ratio ($R = Na^+$ or K^+) and liquid to solid ratio [96–99]. Likewise, the specific parameters of the obtaining process that influence the final properties of the geopolymer are both the curing time and the temperature, as well as the mixing mode and time [96,100,101].

According to several studies [102,103], the compressive strength of geopolymers is positively influenced by the percentage of amorphous phase in the structure. On top of, it was proved that the mechanical properties are influenced by the SiO_2/R_2O ratio, thus, with the increase of R content or with the decrease of Si content, which directly influences the formation of the aluminosilicate network structure, the increase of compression resistance [104]. However, by adding a small amount of sodium silicate to the NaOH solution, the formation of crystals is significantly reduced [105].

2.3.1.   *The influence of the components on the geopolymers characteristics*

The chemical activator or activation solution plays a vital role in initiating the geopolymerization process. Generally, a strongly alkaline medium is required to increase the hydrolysis of aluminosilicate particles present in the raw material, therefore its concentration has a pronounced effect on the mechanical properties of the final product [106]. On the other hand, the dissolution of Si^+ and Al^+ species during geopolymer synthesis depends mainly on the NaOH solution, therefore, the dissolution capacity of silicon and aluminum is controlled, for the most part, by the NaOH concentration and time [107].

Somna et al. [107] studied the compressive strength of fly ash-based geopolymers, hardened at ambient temperature by changing the NaOH concentration from 4.5 M to 16.5 M. It was observed that by increasing the NaOH concentrations from 4.5 to 9.5 M, there is a significant increase in the compressive strength of the samples. However, the increase of NaOH concentrations from 9.5 to 14 M positively influences the mechanical properties of the final product, but at a slower rate. The influence on the compressive strength of the alkaline activator concentration is largely due to the degree of dissolution of Si and Al. Above the molar concentration of 16.5, a decrease in compressive strength is observed due to the excess of hydroxide groups which determines the precipitation of the aluminosilicate gel at early ages.

While in many papers an increase in compressive strength has been reported with an increase in the concentration of the chemical activator, especially sodium hydroxide, some research shows opposite results. For example, a study by J. He et al. [108], on geopolymers, based on red mud or rice straw ash, concluded that a higher concentration of NaOH causes a decrease in compressive strength. This may be due to the high viscosity of the NaOH solution which reduces the solubility of Si and Al, the excessive concentration of OH^- leading to premature precipitation of the gel, as well as a large number of unreacted particles.

In other studies [31,109], the compressive strength of 15 M activated samples was higher than that of 10 M activated samples, due to the high values of the ratio between Na/Al and Na/Si. In turn, they led to the formation of more viscous geopolymer gels that bind unreacted particles and directly contribute to increasing the mechanical properties of the samples.

One of the technological parameters that produce important effects on the characteristics of geopolymers is the quantitative ratio between the liquid component and the solid component (liquid to solid ratio). According to Z. Yahya et al. [31] the value of the ratio are directly proportional to the viscosity of the mixture. Therefore, at a ratio too low the amount of gel formed is insufficient to produce a compact body, while, by increasing the ratio improvement in compressive strength could be produced, but also an increase in setting time (the period required for curing).

According to N. Ariffin et al. increasing the amount of liquid in the mixture also improves the absorption capacity of heavy metals [110]. The close dependence between the degree of dissolution of the mixture and the amount of liquid in it was also highlighted by X-ray diffraction. According to F.A. Prasetya et al. [111] in the case of a geopolymer activated with 3M NaOH solution with different ratios between solid and liquid, a quantitative increase of the specific phase of activation was observed with the increase of the liquid content.

Moreover, the effects are closely related to the ratio between sodium silicate and sodium hydroxide in the liquid component. According to Z. Yahya et al. [31] it has been observed that a concentration too high can accelerate the dissolution reaction of the material rich in silicon and aluminum, but in excess, it can result in the formation of a particles coating that blocks the solution from coming into contact with the undissolved material, accordingly, large particles remain dissolved only at the surface.

To evaluate the influence of calcium content on the compressive strength and setting time of geopolymers, P. Chindaprasirt et al. [112] replaced different percentages of ash with Portland cement type I (CP), calcium hydroxide $(Ca(OH)_2)$ and calcium oxide (CaO). According to the study, the increase in calcium content in the composition causes a sharp decrease in curing time regardless of the percentage of replacement (Figure 2.12).

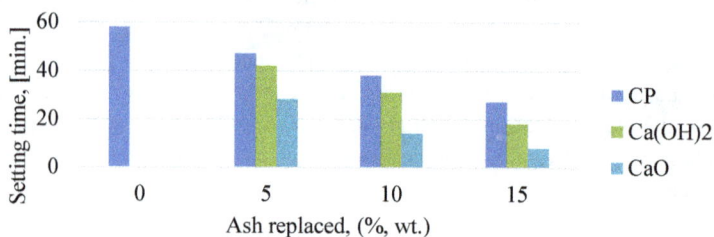

Figure 2.12. Calcium content influence on setting time in geopolymers [109].

According to C. Villa et al. [61], a Na_2SiO_3/NaOH ratio above 1.5 would cause a decrease in compressive strength, as excess silicate causes rapid precipitation of the Al-Si phase from the surface of the particles and prevents the reaction of the phases in the inner layers. However, the results are often contradictory, D. Hardjito et al. [113] obtained the improvement of mechanical properties up to ratios of 2.5. Therefore, the ratio between solid and liquid as well as the ratio between Na_2SiO_3/NaOH must be determined experimentally depending on the source of the raw material.

In addition to the characteristics of the alkaline activator (Na_2SiO_3/NaOH ratio of NaOH solution concentration) and the ratio between solid and liquid, the type of raw material used to make the geopolymer plays an essential role in developing (improving over time by self-healing) mechanical strength properties, durability and microstructure of the resulting material. According to several studies [18,114,115], the particle size distribution during the gel phase has a significant effect on the physical properties and microstructure. In general, pastes with fine particles will acquire, after hardening, superior mechanical properties and dense microstructures [115]. Chindaprasirt et al. [112] reported an improvement in the shrinkage of mortars made from fly ash with a high calcium content 1000% higher than that of Portland cement mortars.

The influence of the elements from the composition on the characteristics of geopolymers was also presented in the study of D.D. Burduhos Nergis et al. [44], were it was confirmed that the compressive strength of fly ash-based geopolymers is closely related to the ratio between the amount of sodium silicate and that of sodium hydroxide in the activator (Figure 2.13), as well as its concentration (Figure 2.14). For the activated geopolymer with a ratio of 2.5 Na_2SiO_3/NaOH in a percentage of 1.5 %, the compressive strength obtained after 7 days was about 40 MPa, for 2 % about 70 MPa, and 2.5 % of about 55 MPa.

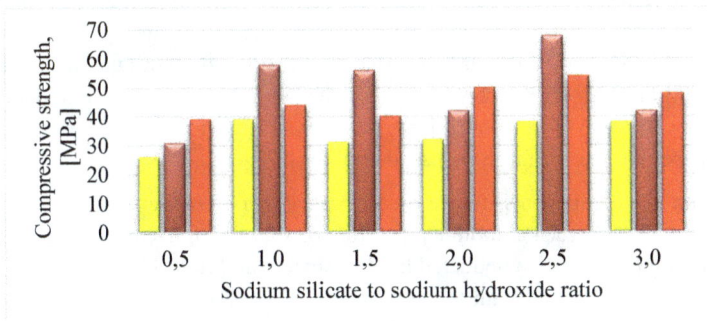

Figure 2.13. The influence of Na2SiO3 to NaOH ratio on compressive strength [44].

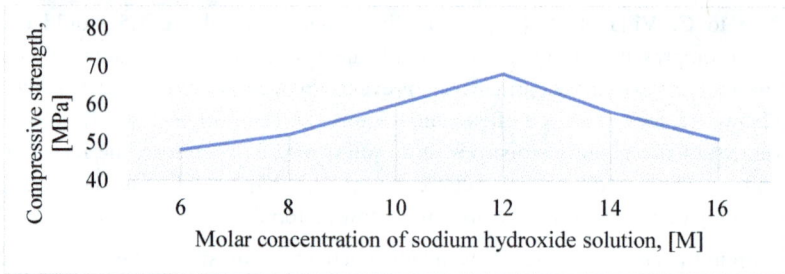

Figure 2.14. The influence of NaOH solution concentration on compressive strength [44].

In order to obtain an efficient synthesis, a balance must be obtained between the essential elements of geopolymerization, namely SiO_2, Al_2O_3, Na_2O and CaO. This balance can be achieved most quickly by the addition of additives rich in the necessary chemical elements, such as calcium hydroxide ($Ca(OH)_2$), alumina hydroxide ($Al(OH)_3$), nanosilica oxide, aluminum nano-oxide etc. [116,117]. Furthermore, the addition of different types of fibers, polymeric resins, super-plasticizers and nanomaterials significantly improves the properties of geopolymers, especially mechanical properties, such as compressive strength or bending strength [44,92,118–120].

2.3.2. The effect of obtaining process parameters

Conventional methods of making geopolymers involve mixing the solid component, rich in aluminum and silicon, with a liquid component, a strongly alkaline solution. There are multiple studies related to the ideal way of mixing. However, the most common method involves mixing the solid components together, if there are several, as well as the liquid ones together, followed by mixing the liquid component with the solid one [121,122]. Another method is to mix the material rich in silicon and aluminum with the sodium silicate solution, and after a few minutes, the NaOH solution is introduced [123]. However, the geopolymers obtained by the second method exhibit lower mechanical strength properties, except for those based on fly ash.

After mixing the components, the resulting paste (geopolymeric binder) is poured into molds, where it will harden at room temperature or slightly high temperature. At the same time, according to the study conducted by Z. Yahya et al. [109], if after filling the molds, the surface in direct contact with the atmosphere is covered with a plastic thin sheet the water evaporation rate from the mixture is reduced, therefore, the geopolymer will show a smaller number of cracks.

In the case of clay, an additional supply of water will be needed for the same amount of solid material. Geopolymer clay paste is much more sticky than fly ash paste [124]. This is because clay has a layered structure that reduces workability (*definition:* "Workability is the set of properties that allows homogeneity (ie non-separation of components) during the handling, transport, compaction and finishing of fresh concrete. Workability can be characterized and by the property of the concrete to fill the form easily and to embed the reinforcements during the casting" [125]). Thus, the low workability of the geopolymer paste can lead to compaction difficulties, followed by a final porous structure and low mechanical properties. In the case of flying ash-based geopolymers, the high mechanical properties may also be due to the spherical shape of the particles involving a low frictional force [123].

The most important stage of the formation process of geopolymers is the heat treatment of curing (drying) because it influences the rate of polymerization and solidification [89,90]. Although the geopolymerization reaction can also take place at ambient temperature, in most studies a temperature between 60 and 90 °C was used. The use of too low drying temperature (20 ÷ 60 °C) produces a slow evolution of the geopolymerization reaction [126,127] and, therefore, a gradual improvement of the mechanical properties. However, the use of too high drying temperature (80 ÷ 100 °C) produces a rapid dissolution of the surface of the particles in the base material, but also rapid evaporation of water [128–130]. S. B. Amar et al. [131] evaluated the dependence between the drying temperature and the concentration of the sodium hydroxide solution on the compressive strength of the geopolymers (Figure 2.15). According to the study, the increase of the drying temperature from 30 to 80 °C produced positive effects on the compressive strength as a result of supporting the endothermic geopolymerization reaction. However, above 80 °C, the rapid evaporation of water causes cracks and, consequently, a decrease in compressive strength.

According to the study conducted by Mo. Bing-Hui et al. [132], the increase of drying temperature accelerates dissolution and polymerization in the case of metakaolin. Geopolymers made at drying temperatures between 60 °C and 80 °C, show an increase in mechanical properties, compared to those made at ambient temperature. However, above 80 °C (80 ÷ 100) °C, the increase in temperature causes the mechanical properties to decrease due to a large number of cracks. Another study showed that a geopolymer dried at room temperature for one month has a lower compressive strength than the one dried at 80 °C for 24 hours. According to the study conducted by Palomo et al. [133], on power plant ash-based geopolymers, it was shown that the compressive strength of the hardened geopolymer at 85 °C for 24 hours was higher than that of the heat-treated sample at 65 °C. However, increasing the curing temperature from 65 °C to 80 °C accelerates the increase in compressive strength of the samples for the first 28 days; however, after this period the mechanical properties begin to deteriorate, compared to those of the samples hardened at ambient temperature, according to the study presented by Rovnanik [134].

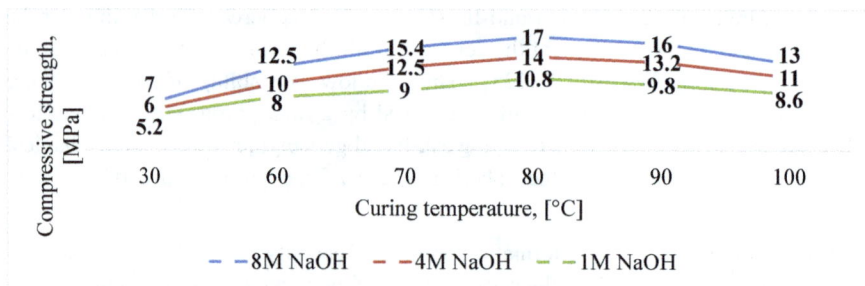

Figure 2.15. The effects of curing temperature and activator concentration on compressive strength [131].

Except from the temperature at which the geopolymer hardens (curing stage), another parameter that influences the drying process is the maintaining time, because a too short period produces minimal effects on the geopolymerization reaction, and a long period is not efficient from the economic point of view [135,136]. However, there is no optimal value for temperature and maintaining time, for all types of geopolymers, therefore they must be determined experimentally.

2.4. Geopolymers characteristics

2.4.1. Effect of acid attack on geopolymers

The degradation of geopolymers during exposure to acidic environments is based on the depolymerization reaction of the aluminosilicate network (Figure 2.16) [122]. The degree and rate of deterioration of the structure are closely related, in general, to the concentration of the acidic environment and the time of exposure. According to the study conducted by Davidovits et al. [9], the mass loss of kaolin-based geopolymer samples immersed in 5% H_2SO_4 solution for 30 days was only of approximatively 7 %.

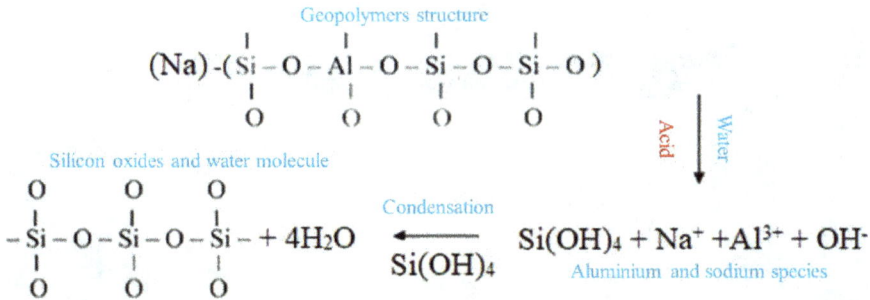

Figure 2.16. Schematic representation of the depolymerization reactions of geopolymers in the acidic environment [122].

Furthermore, it was shown that after 3 months of exposure in HNO_3 of coal-ash based geopolymers, their microstructure remains relatively unaffected. However, Temuujin et al. [137] concluded that the acid and alkaline resistance of geopolymers based on thermal power plant ash depends mainly on its mineralogical composition. This statement is based on the difference in solubility between Al, Si and Fe ions.

Another study [122] analyzed the decrease in compressive strength of fly ash-based geopolymers exposed in three types of acidic environments for different periods. According to the study, by comparing the effects produced by the three corrosive environments, the HNO_3, H_2SO_4 and HCl media, hydrochloric acid showed the most aggressive attack, despite the exposure period. Moreover, the effect of structure reinforcing with fine aggregates (class 0/4 sand) in the proportion of 0%, 75% and 50% of the solid component was evaluated. Analyzing the data from the point of view of the decrease in compressive strength as a function of the exposure period, it can be seen that the sample with the highest amount of sand showed the lowest effects. This result may be due to the reduction in the rate of acid penetration due to the increased compactness of the samples because the sand particles cannot be depolymerized functioning as protective shields for the sample matrix. As can be seen from the 3D profile of the sample surface (Figure 2.17), during the exposure period several cracks appear that start from the surface and advance towards the inside of the sample.

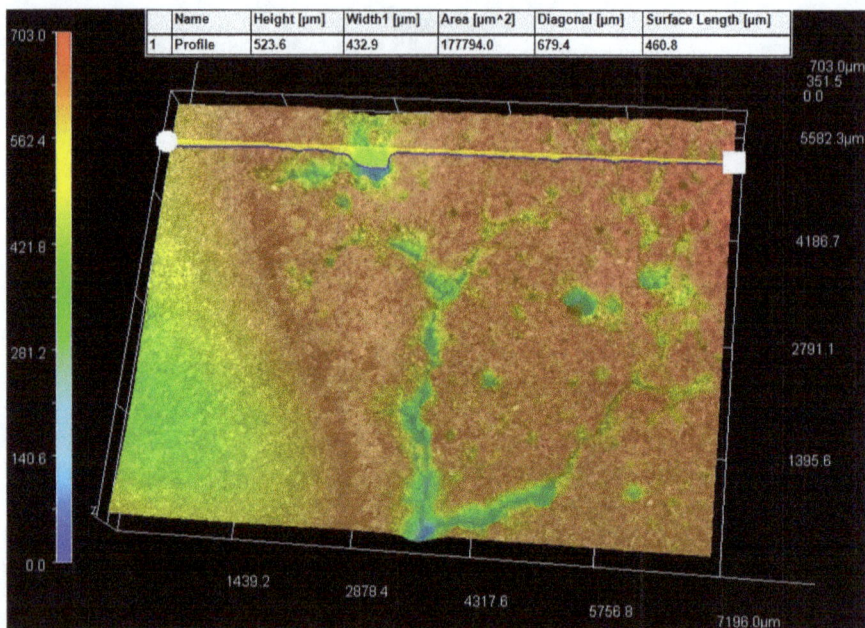

Name	Height [μm]	Width1 [μm]	Area [μm^2]	Diagonal [μm]	Surface Length [μm]
1 Profile	523.6	432.9	177794.0	679.4	460.8

Figure 2.17. 3D Profile of the acid exposed samples [122].

In the study of Hardjito et al. [138], a 20 % reduction in the compressive strength of ash-based geopolymers exposed to a 0.5 % H_2SO_4 solution for 12 months was measured. It was also found that with the increase of the concentration from 0.5 % to 1 % and 2 %, respectively, the decrease was 52 % and 65 %, respectively. The significant loss of mechanical properties is mainly due to the accelerated degradation of the matrix, as opposed to that of the aggregate. Ariffin et al. [139] analyzed the effect of the 2 % sulfuric acid solution on the compressive strength and mass loss of a geopolymer based on thermal power plant ash and palm oil ash compared to that of a Portland cement-based concrete. After 18 months the mass loss of the geopolymer was close to 8 % while that of conventional concrete was close to 20 %. In terms of compressive strength, the geopolymer showed a decrease of 35 %, but Portland cement-based concrete suffered severe damage resulting in a decrease of 68 %. They concluded that the C-S-H group (hydrated calcium silicate), on which Portland cement-based concretes are based, has lower acid resistance compared to the N-A-S-H group (sodium silico-aluminate hydrate), specific for geopolymers.

N. Rajamane et al. [140] measured a decrease in compressive strength of between 2 % and 29 % in the case of geopolymers exposed to attack by a 5 % Na_2SO_4 and 5 % $MgSO_4$ solution. Under the same exposure conditions, conventional concrete samples showed a decrease between 9 % and 38 %. According to the study, the negative effects on the mechanical properties are due to the formation of gypsum and expansive shrinkage that can cause the concrete to expand, crack and break. Therefore, the effects are lower on geopolymers due to the lower content in their structure of $Ca(OH)_2$ and mono-sulfoaluminate.

2.4.2. Dimensional stability and durability

The characteristics of the alkaline activator and the hardening regime also have a direct effect on the dimensional stability and durability of geopolymers. Ridtirud et al. [141] investigated the effect of the concentration of NaOH and sodium silicate on the shrinkage of Class C ash based geopolymer. According to the study, geopolymers dried at 40 °C for 24 h showed an increase in shrinkage directly proportional to the increase in the concentration of sodium hydroxide solution in the activator. On the other hand, the increase in the ratio between sodium silicate and sodium hydroxide produces geopolymers with significantly lower shrinkage values, due to the high ratio between silicon oxide and aluminum oxide which produces a rapid condensation of the geopolymer. Moreover, the effect of curing temperature on the contraction of geopolymer mortars was also studied. The results show that samples exposed to a higher curing temperature, 60 °C, suffer a lower shrinkage than those cured at ambient temperature.

According to the study by N. Değirmenci [142], geopolymers based on thermal power plant ash or blast furnace slag showed no visible damage after being subjected to 25 freeze-thaw cycles. However, from the weight changes point of view, the samples showed decreases between 5 % and 22 %. The effect produced by 28 freeze-thaw cycles on compressive strength is presented in the study published by S. Pilehvar et al. [143]. According to this, the decrease by 1 % of the compressive strength is due to the expansion of water, during frosting, from the capillary pores that produce the appearance of microcracks in the sample matrix.

2.4.3. Permeability

To study the gas permeability, Bernal et al. [144] evaluated the decrease in compressive strength of some geopolymers based on metakaolin and slag depending on the carbonation period. Accordingly, it was observed that the decrease in compressive strength is directly proportional to the exposure period. Therefore, the evaluated samples allow the advancement of the gas with 3.0 ± 0.2 % CO_2 which produces negative effects on the mechanical properties.

In another study, Olivia et al. [146] evaluated the water permeability of a hardened geopolymer at 60 °C for 24 h, the values obtained were lower than the Portland cement samples due to the density of the paste and the interconnectivity of the small pores. They also pointed out that the water/solid geopolymer ratio was the most significant parameter affecting the properties of the geopolymer. Bondar et al. [147] developed a comparative study between the oxygen and chlorine permeability of some Portland cement-based concretes and geopolymers. According to the study, after a hardening period of 90 days, the permeability of the geopolymer is 10-35 % lower for oxygen but much higher for chlorine, compared to that of Portland cement samples. This is due to the very high concentration of alkaline ions in the remaining solution in the pores which increases the thermal conductivity. However, the amount of fluid in the pores increases depending on the age of the samples.

Despite the many advantages presented by geopolymers over conventional materials, worldwide there are still many challenges to be overcome for the successful application at industrial level of geopolymers. These restrains are related to the lack of standards and evaluations in real environmental conditions of long-term geopolymers. Moreover, geopolymers that use certain natural minerals (kaolin, clay etc.) as a source of raw materials are expensive and produce negative effects on the mining areas. By comparison, this research addresses the obtaining of geopolymers, without Portland cement, by recovering mineral waste (power plant ash and glass powder), thus producing positive effects on the environment, by greening storage areas or dumps.

E. Spataru [145] performed water absorption tests, according to ASTM C642-13 standard, on geopolymers obtained from 85% and 95% granular blast furnace slag (ZG) mixed with ash (CPG) (15%, respectively 5 %) and/or red mud (NR) (15% and 5% respectively) (Figure 2.18). According to the study, it was observed that the degree of absorption of geopolymers is influenced by the type of raw material used as well as the curing temperature (20 °C, 40 °C and 60 °C). For example, the water absorption capacity of red mud samples cured at 60 °C was approximately 25 % lower than that of ash samples.

Although worldwide, geopolymers are of great interest for research in the field of materials engineering, in many countries the concerns for these materials are quite low. Although, there are large quantities of mineral waste (power plant ash, red sludge, mining tailings, etc.) with potential for geopolymerization that could be exploited through this technology.

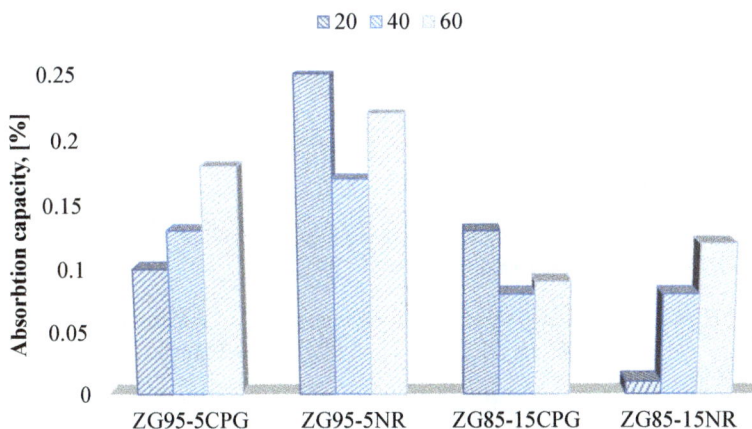

Figure 2.18. Water absorption of 7 days geopolymers [145].

Considering the diversity of raw material sources, the different methods of obtaining and the parameters with major influence on the properties of geopolymers, presented in the literature, it was concluded that it is necessary to evaluate and experimentally determine the optimal parameters for the obtaining process. Accordingly, the following parameters must be studied and customized according to the raw materials: the characteristics of the raw material, the ratio between the alkaline activator components, the ratio between solid and liquid, the curing time, the curing temperature and the possibility of adding reinforcing particles.

References

1.   Davidovits, J. Geopolymers - Inorganic polymeric new materials. *J. Therm. Anal.* 1991, *37*, 1633–1656. https://doi.org/10.1007/BF01912193

2.   Bell, J.L.; Driemeyer, P.E.; Kriven, W.M. Formation of Ceramics from Metakaolin-Based Geopolymers. Part II: K-Based Geopolymer. *J. Am. Ceram. Soc.* 2009, *92*, 607–615. https://doi.org/10.1111/j.1551-2916.2008.02922.x

3.   Provis, J.L.; van Deventer, J.S.J. Geopolymerisation kinetics. 2. Reaction kinetic modelling. *Chem. Eng. Sci.* 2007, *62*, 2318–2329. https://doi.org/10.1016/j.ces.2007.01.028

4.   Provis, J.L.; van Deventer, J.S.J. Geopolymerisation kinetics. 1. In situ energy-dispersive X-ray diffractometry. *Chem. Eng. Sci.* 2007, *62*, 2309–2317. https://doi.org/10.1016/j.ces.2007.01.027

5.   Burduhos Nergis, D.D.; Abdullah, M.M.A.B.; Vizureanu, P.; Mohd Tahir, M.F.M. Geopolymers and Their Uses: Review. In Proceedings of the IOP Conference Series: Materials Science and Engineering; 2018; Vol. 374.

6.   Davidovits, J. *30 Years of Successes and Failures in Geopolymer Applications. Market Trends and Potential Breakthroughs.*

7.   Xu, H.; Van Deventer, J.S.J. The geopolymerisation of alumino-silicate minerals. *Int. J. Miner. Process.* 2000, *59*, 247–266. https://doi.org/10.1016/S0301-7516(99)00074-5

8.   Xu, H.; Deventer, J.S.J. Van Geopolymerisation of multiple minerals. *Miner. Eng.* 2002, *15*, 1131–1139. https://doi.org/10.1016/s0892-6875(02)00255-8

9.   Davidovits, J., Geopolymer, Green Chemistry and Sustainable Development Solutions ... - Google Books, available online, (accessed on Feb 2, 2020).

10.   Mackenzie, K.J.D.; Welter, M. Geopolymer (aluminosilicate) composites: Synthesis, properties and applications. In *Advances in Ceramic Matrix Composites*; Elsevier Ltd., 2014; pp. 445–470 ISBN 9780857091208.

11.   Siyal, A.A.; Azizli, K.A.; Man, Z.; Ullah, H. Effects of Parameters on the Setting Time of Fly Ash Based Geopolymers Using Taguchi Method. In Proceedings of the Procedia Engineering; Elsevier Ltd, 2016; Vol. 148, pp. 302–307.

12.   Gao, X.; Yu, Q.L.; Brouwers, H.J.H. Reaction kinetics, gel character and strength of ambient temperature cured alkali activated slag-fly ash blends. *Constr. Build. Mater.* 2015, *80*, 105–115. https://doi.org/10.1016/j.conbuildmat.2015.01.065

13.   Pelisser, F.; Silva, B. V.; Menger, M.H.; Frasson, B.J.; Keller, T.A.; Torii, A.J.; Lopez, R.H. Structural analysis of composite metakaolin-based geopolymer concrete. *Rev. IBRACON Estruturas e Mater.* 2018, *11*, 535–543. https://doi.org/10.1590/s1983-41952018000300006

14.   Barbosa, V.F.F.; MacKenzie, K.J.D. Synthesis and thermal behaviour of potassium sialate geopolymers. *Mater. Lett.* 2003, *57*, 1477–1482. https://doi.org/10.1016/S0167-577X(02)01009-1

15.   Koleżyński, A.; Król, M.; Żychowicz, M. The structure of geopolymers –
Theoretical studies. *J. Mol. Struct.* 2018. https://doi.org/10.1016/j.molstruc.2018.03.033.

16.   Osada, T.; Nakao, W.; Takahashi, K.; Ando, K. Self-crack-healing behavior in
ceramic matrix composites. In *Advances in Ceramic Matrix Composites*; Elsevier Ltd.,
2014; pp. 410–441 ISBN 9780857091208.

17.   Ma, H.; Qian, S.; Li, V.C. Influence of fly ash type on mechanical properties and
self-healing behavior of Engineered Cementitious Composite (ECC). In Proceedings of
the Proceedings of the 9th International Conference on Fracture Mechanics of Concrete
and Concrete Structures; IA-FraMCoS, 2016.

18.   Zain, H.; Abdullah, M.M.A.B.; Hussin, K.; Ariffin, N.; Bayuaji, R. Review on
Various Types of Geopolymer Materials with the Environmental Impact Assessment.
MATEC Web Conf. 2017, *97*, 01021. https://doi.org/10.1051/matecconf/20179701021

19.   Davidovits, J. Geopolymers Based on Natural and Synthetic Metakaolin a Critical
Review. In; 2018; pp. 201–214.

20.   Yu, J.; Yang, Y.; Chen, W.; Xu, D.; Guo, H.; Li, K.; Liu, H. The synthesis and
application of zeolitic material from fly ash by one-pot method at low temperature. *Green
Energy Environ.* 2016, *1*, 166–171. https://doi.org/10.1016/j.gee.2016.07.002

21.   Riahi, S.; Nazari, A.; Zaarei, D.; Khalaj, G.; Bohlooli, H.; Kaykha, M.M.
Compressive strength of ash-based geopolymers at early ages designed by Taguchi
method. *Mater. Des.* 2012, *37*, 443–449. https://doi.org/10.1016/j.matdes.2012.01.030

22.   Nazari, A.; Bagheri, A.; Riahi, S. Properties of geopolymer with seeded fly ash
and rice husk bark ash. *Mater. Sci. Eng. A* 2011, *528*, 7395–7401.
https://doi.org/10.1016/j.msea.2011.06.027

23.   Chen, L.; Wang, Z.; Wang, Y.; Feng, J. Preparation and properties of alkali
activated metakaolin-based geopolymer. *Materials (Basel).* 2016, *9*.
https://doi.org/10.3390/ma9090767

24.   Karakoç, M.B.; Türkmen, I.; Maraş, M.M.; Kantarci, F.; Demirbola, R.; Uſur
Toprak, M. Mechanical properties and setting time of ferrochrome slag based
geopolymer paste and mortar. *Constr. Build. Mater.* 2014, *72*, 283–292.
https://doi.org/10.1016/j.conbuildmat.2014.09.021

25.   Abdullah, M.M.A.B.; Ming, L.Y.; Yong, H.C.; Tahir, M.F.M. Clay-Based
Materials in Geopolymer Technology. In *Cement Based Materials*; InTech, 2018.

26.   Yunsheng, Z.; Wei, S.; Zongjin, L. Composition design and microstructural characterization of calcined kaolin-based geopolymer cement. *Appl. Clay Sci.* 2010, *47*, 271–275. https://doi.org/10.1016/j.clay.2009.11.002

27.   Yang, L.; Qian, X.; Yuan, P.; Bai, H.; Miki, T.; Men, F.; Li, H.; Nagasaka, T. Green synthesis of zeolite 4A using fly ash fused with synergism of NaOH and Na2CO3. *J. Clean. Prod.* 2019, *212*, 250–260. https://doi.org/10.1016/j.jclepro.2018.11.259

28.   Heah, C.Y.; Kamarudin, H.; Bakri, A.M.M. Al; Binhussain, M.; Musa, L.; Nizar, I.K.; Ghazali, C.M.R.; Liew, Y.M. Curing Behavior on Kaolin-Based Geopolymers. *Adv. Mater. Res.* 2012, *548*, 42–47. https://doi.org/10.4028/www.scientific.net/amr.548.42

29.   Morales Herrera, A.; Moreira Palacios, M. Propagación de Erato polymnioides, en combinaciones de sustratos, reguladores de crecimiento y agrupación de plántulas. *Agron. Costarric.* 2019. https://doi.org/10.15517/rac.v44i1.40019

30.   Varga, G. The structure of kaolinite and metakaolinite. *Epa. - J. Silic. Based Compos. Mater.* 2007, *59*, 6–9. https://doi.org/10.14382/epitoanyag-jsbcm.2007.2

31.   Yahya, Z.; Abdullah, M.M.A.B.; Ramli, N.M.; Burduhos-Nergis, D.D.; Razak, R.A. Influence of Kaolin in Fly Ash Based Geopolymer Concrete: Destructive and Non-Destructive Testing. *IOP Conf. Ser. Mater. Sci. Eng.* 2018, *374*, 12068. https://doi.org/10.1088/1757-899x/374/1/012068

32.   Harekrushna, S.; Subash, C.M.; Santosh, K.S.; Ananta, P.; Himanshu, S.M. Progress of Red Mud Utilization: An Overview. *Am. Chem. Sci. J.* 2014, *4*, 255–279.

33.   Dwivedi, A.; Jain, M.K. Fly ash – waste management and overview : A Review . *Recent Res. Sci. Technol.* 2014, *6*, 30–35.

34.   Ye, J.; Zubair, M.; Wang, S.; Cai, Y.; Zhang, P. Power production waste. *Water Environ. Res.* 2019, *91*, 1091–1096.

35.   Xiong, X.; Liu, X.; Yu, I.K.M.; Wang, L.; Zhou, J.; Sun, X.; Rinklebe, J.; Shaheen, S.M.; Ok, Y.S.; Lin, Z.; et al. Potentially toxic elements in solid waste streams: Fate and management approaches. *Environ. Pollut.* 2019, *253*, 680–707.

36.   Coal utilization - Coal combustion | Britannica Available online: https://www.britannica.com/topic/coal-utilization-122944/Coal-combustion (accessed on Feb 8, 2020).

37.   Hurskainen, M.; Vainikka, P. Technology options for large-scale solid-fuel combustion. In *Fuel Flexible Energy Generation: Solid, Liquid and Gaseous Fuels*; Elsevier Inc., 2016; pp. 177–199 ISBN 9781782423997.

38.   Bricl, M. Cleaning of flue gases in thermal power plants čiščenje dimnih plinov v termoenergetskih postrojenjih; 2016; Vol. 9;.

39.   Karakashev, D. Zhang, Y., BioEnergy and BioChemicals Production from Biomass and Residual Resources - Google Books, available online: (accessed on Feb 13, 2020).

40.   Argiz, C.; Sanjuán, M.Á.; Menéndez, E. Coal Bottom Ash for Portland Cement Production. *Adv. Mater. Sci. Eng.* 2017, *2017*. https://doi.org/10.1155/2017/6068286

41.   Rudd, H.L. Chemicals in the environment. *Calif. Med.* 1970, *113*, 27–32. https://doi.org/10.1016/b978-0-12-804422-3.00002-x

42.   Komljenović, M.; Baščarević, Z.; Bradić, V. Mechanical and microstructural properties of alkali-activated fly ash geopolymers. *J. Hazard. Mater.* 2010, *181*, 35–42. https://doi.org/10.1016/j.jhazmat.2010.04.064

43.   Sathonsaowaphak, A.; Chindaprasirt, P.; Pimraksa, K. Workability and strength of lignite bottom ash geopolymer mortar. *J. Hazard. Mater.* 2009, *168*, 44–50. https://doi.org/10.1016/j.jhazmat.2009.01.120

44.   Doru Dumitru, N.B.; Al Bakri Abdullah, M.M.; Petrică, V. The effect of fly ash/alkaline activator ratio in class F fly ash based geopolymers. *Eur. J. Mater. Sci. Eng.* 2017, *2*, 111–118.

45.   Onprom, P.; Chaimoon, K.; Cheerarot, R. Influence of Bottom Ash Replacements as Fine Aggregate on the Property of Cellular Concrete with Various Foam Contents. *Adv. Mater. Sci. Eng.* 2015, *2015*. https://doi.org/10.1155/2015/381704

46.   Wattimena, O.K.; Antoni; Hardjito, D. A review on the effect of fly ash characteristics and their variations on the synthesis of fly ash based geopolymer. In Proceedings of the AIP Conference Proceedings; American Institute of Physics Inc., 2017; Vol. 1887.

47.   Thomas, M., Optimizing the Use of Fly Ash in Concrete. https://www.cement.org/docs/default-source/fc_concrete_technology/is548-optimizing-the-use-of-fly-ash-concrete.pdf (accessed on Feb 13, 2020).

48.   Vassilev, S. V.; Menendez, R.; Alvarez, D.; Diaz-Somoano, M.; Martinez-Tarazona, M.R. Phase-mineral and chemical composition of coal fly ashes as a basis for their multicomponent utilization. 1. Characterization of feed coals and fly ashes. *Fuel* 2003, *82*, 1793–1811. https://doi.org/10.1016/S0016-2361(03)00123-6

49.   Moghal, A.A.B. Geotechnical and physico-chemical characterization of low lime fly ashes. *Adv. Mater. Sci. Eng.* 2013, *2013*. https://doi.org/10.1155/2013/674306

50.   ASTM C 618-12a Coal Fly Ash and Raw or Calcined Natural Pozzolan for Use in Concrete. *ASTM Int.* 2012, 1–5.

51.   Khairul, M.A.; Zanganeh, J.; Moghtaderi, B. The composition, recycling and utilisation of Bayer red mud. *Resour. Conserv. Recycl.* 2019, *141*, 483–498.

52.   Pashias, N., Doger, V., Essential Readings in Light Metals, Volume 1, Alumina and Bauxite - Google books, available online, (accessed on Feb 13, 2020).

53.   Kang, S.-P.; Kwon, S.-J. Effects of red mud and Alkali-Activated Slag Cement on efflorescence in cement mortar. *Constr. Build. Mater.* 2017, *133*, 459–467. https://doi.org/10.1016/j.conbuildmat.2016.12.123

54.   Bayer Process: Manufacturing Of Alumina | Making of Alumina Available online: https://www.worldofchemicals.com/591/chemistry-articles/manufacturing-of-alumina-through-bayer-process.html (accessed on Feb 13, 2020).

55.   Pepper, R.A.; Couperthwaite, S.J.; Millar, G.J. Comprehensive examination of acid leaching behaviour of mineral phases from red mud: Recovery of Fe, Al, Ti, and Si. *Miner. Eng.* 2016, *99*, 8–18. https://doi.org/10.1016/j.mineng.2016.09.012

56.   Bernardo-Maestro, B.; López-Arbeloa, F.; Pérez-Pariente, J.; Gómez-Hortigüela, L. Comparison of the structure-directing effect of ephedrine and pseudoephedrine during crystallization of nanoporous aluminophosphates. *Microporous Mesoporous Mater.* 2017, *254*, 211–224. https://doi.org/10.1016/j.micromeso.2017.04.008

57.   Olszewska, J.P.; Heal, K. V.; Winfield, I.J.; Eades, L.J.; Spears, B.M. Assessing the role of bed sediments in the persistence of red mud pollution in a shallow lake (Kinghorn Loch, UK). *Water Res.* 2017, *123*, 569–577. https://doi.org/10.1016/j.watres.2017.07.009

58.   Sani, N.A.M.; Man, Z.; Shamsuddin, R.M.; Azizli, K.A.; Shaari, K.Z.K. Determination of Excess Sodium Hydroxide in Geopolymer by Volumetric Analysis. In Proceedings of the Procedia Engineering; Elsevier Ltd, 2016; Vol. 148, pp. 298–301.

59.   Khale, D.; Chaudhary, R. Mechanism of geopolymerization and factors influencing its development: A review. *J. Mater. Sci.* 2007, *42*, 729–746. https://doi.org/10.1007/s10853-006-0401-4

60.   Merabtene, M.; Kacimi, L.; Clastres, P. Elaboration of geopolymer binders from poor kaolin and dam sludge waste. *Heliyon* 2019, *5*, e01938. https://doi.org/10.1016/j.heliyon.2019.e01938

61.   Villa, C.; Pecina, E.T.; Torres, R.; Gómez, L. Geopolymer synthesis using alkaline activation of natural zeolite. *Constr. Build. Mater.* 2010, *24*, 2084–2090. https://doi.org/10.1016/j.conbuildmat.2010.04.052

62.   Torres-Carrasco, M.; Palomo, J.G.; Puertas, F. Sodium silicate solutions from dissolution of glasswastes. Statistical analysis. *Mater. Construcción* 2014, *64*, e014. https://doi.org/10.3989/mc.2014.05213

63.   Qiu, T.; Kuang, J.; Shi, F. Effect of alkali on the geopolymer strength. In Proceedings of the Advanced Materials Research; Trans Tech Publications Ltd, 2011; Vol. 168–170, pp. 1827–1832.

64.   Oleiwi, S.M.; Algın, Z.; Nassani, D.E.; Mermerdaş, K. Multi-Objective Optimization of Alkali Activator Agents for FA- and GGBFS-Based Geopolymer Lightweight Mortars. *Arab. J. Sci. Eng.* 2018, *43*, 5333–5347. https://doi.org/10.1007/s13369-018-3170-x

65.   Abdul Rahim, R.H.; Rahmiati, T.; Azizli, K.A.; Man, Z.; Nuruddin, M.F.; Ismail, L. Comparison of using NaOH and KOH activated fly ash-based geopolymer on the mechanical properties. In Proceedings of the Materials Science Forum; Trans Tech Publications Ltd, 2015; Vol. 803, pp. 179–184.

66.   Kabir, S.M.A.; Alengaram, U.J.; Jumaat, M.Z.; Sharmin, A.; Islam, A. Influence of molarity and chemical composition on the development of compressive strength in POFA based geopolymer mortar. *Adv. Mater. Sci. Eng.* 2015, *2015*. https://doi.org/10.1155/2015/647071

67.   Mangat, P.; Lambert, P. Sustainability of alkali-activated cementitious materials and geopolymers. In *Sustainability of Construction Materials*; Elsevier, 2016; pp. 459–476.

68.   Ridzuan, A.R.M.; Khairulniza, A.A.; Arshad, M.F. Effect of sodium silicate types on the high calcium geopolymer concrete. In Proceedings of the Materials Science Forum; Trans Tech Publications Ltd, 2015; Vol. 803, pp. 185–193.

69.   Panagiotopoulou, C.; Kontori, E.; Perraki, T.; Kakali, G. Dissolution of aluminosilicate minerals and by-products in alkaline media. *J. Mater. Sci.* 2006, *42*, 2967–2973. https://doi.org/10.1007/s10853-006-0531-8

70.   Xu, H.; Van Deventer, J.S.J. Effect of source materials on geopolymerization. *Ind. Eng. Chem. Res.* 2003, *42*, 1698–1706. https://doi.org/10.1021/ie0206958

71.   Panagiotopoulou, C.; Kontori, E.; Perraki, T.; Kakali, G. Dissolution of aluminosilicate minerals and by-products in alkaline media. *J. Mater. Sci.* 2007, *42*, 2967–2973. https://doi.org/10.1007/s10853-006-0531-8

72.   Rahier, H.; Wastiels, J.; Biesemans, M.; Willlem, R.; Van Assche, G.; Van Mele, B. Reaction mechanism, kinetics and high temperature transformations of geopolymers. *J. Mater. Sci.* 2007, *42*, 2982–2996. https://doi.org/10.1007/s10853-006-0568-8

73.   Lizcano, M.; Kim, H.S.; Basu, S.; Radovic, M. Mechanical properties of sodium and potassium activated metakaolin-based geopolymers. *J. Mater. Sci.* 2011, *47*, 2607–2616. https://doi.org/10.1007/s10853-011-6085-4

74.   Steveson, M.; Sagoe-Crentsil, K. Relationships between composition, structure and strength of inorganic polymers. *J. Mater. Sci.* 2005, *40*, 2023–2036. https://doi.org/10.1007/s10853-005-1226-2

75.   Steveson, M. Relationships between composition , structure. *Mater. Sci.* 2005, *40*, 4247–4259. https://doi.org/10.1007/s10853-005-1226-2

76.   Nergis, D.D.B.; Al Bakri Abdullah, M.M.; Sandu, A.V.; Vizureanu, P. XRD and TG-DTA study of new alkali activated materials based on fly ash with sand and glass powder. *Materials (Basel).* 2020, *13*. https://doi.org/10.3390/ma13020343

77.   Burduhos Nergis, D.D.; Vizureanu, P.; Ardelean, I.; Sandu, A.V.; Corbu, O.C.; Matei, E. Revealing the Influence of Microparticles on Geopolymers' Synthesis and Porosity. *Materials (Basel).* 2020, *13*, 3211. https://doi.org/10.3390/ma13143211

78.   Duxson, P.; Fernández-Jiménez, A.; Provis, J.L.; Lukey, G.C.; Palomo, A.; van Deventer, J.S.J. Geopolymer technology: the current state of the art. *J. Mater. Sci.* 2006, *42*, 2917–2933. https://doi.org/10.1007/s10853-006-0637-z

79.   Provis, J.L. Activating solution chemistry for geopolymers. In *Geopolymers: Structures, Processing, Properties and Industrial Applications*; Elsevier Ltd., 2009; pp. 50–71 ISBN 9781845694494.

80.   Kong, D.L.Y.; Sanjayan, J.G.; Sagoe-Crentsil, K. Comparative performance of geopolymers made with metakaolin and fly ash after exposure to elevated temperatures. *Cem. Concr. Res.* 2007, *37*, 1583–1589. https://doi.org/10.1016/j.cemconres.2007.08.021

81.   Kong, D.L.Y.; Sanjayan, J.G.; Sagoe-Crentsil, K. Factors affecting the performance of metakaolin geopolymers exposed to elevated temperatures. *J. Mater. Sci.* 2008, *43*, 824–831. https://doi.org/10.1007/s10853-007-2205-6

82.   Barbosa, V.F.F.; MacKenzie, K.J.D.; Thaumaturgo, C. Synthesis and characterisation of materials based on inorganic polymers of alumina and silica: Sodium polysialate polymers. *Int. J. Inorg. Mater.* 2000, *2*, 309–317. https://doi.org/10.1016/S1466-6049(00)00041-6

83.   Manufacture of Soluble Silicates, available online: https://www.cees-silicates.org/index.php/manufacture (accessed on Feb 16, 2020).

84.   Phair, J.W.; Van Deventer, J.S.J. Effect of silicate activator pH on the leaching and material characteristics of waste-based inorganic polymers. *Miner. Eng.* 2001, *14*, 289–304. https://doi.org/10.1016/S0892-6875(01)00002-4

85.   Garcia-Lodeiro, I.; Palomo, A.; Fernández-Jiménez, A. An overview of the chemistry of alkali-activated cement-based binders. In *Handbook of Alkali-Activated Cements, Mortars and Concretes*; Elsevier Inc., 2015; pp. 19–47 ISBN 9781782422884.

86.   Zhang, B.; MacKenzie, K.J.D.; Brown, I.W.M. Crystalline phase formation in metakaolinite geopolymers activated with NaOH and sodium silicate. *J. Mater. Sci.* 2009, *44*, 4668–4676. https://doi.org/10.1007/s10853-009-3715-1

87.   Ismail, I.; Bernal, S.A.; Provis, J.L.; Hamdan, S.; Van Deventer, J.S.J. Microstructural changes in alkali activated fly ash/slag geopolymers with sulfate exposure. *Mater. Struct. Constr.* 2013, *46*, 361–373. https://doi.org/10.1617/s11527-012-9906-2

88.   Production process - silmaco.com Available online: http://www.silmaco.com /production-process (accessed on Feb 16, 2020).

89.   Singh, J.P.; Bansal, N.P.; Kriven, W.M.; American Ceramic Society. Meeting (106th : 2004 : Indianapolis, I..; Ceramic-Matrix Composites Symposium (2004 : Indianapolis, I.. *Microstructural Characterization of Metakaolin-Based Geopolymers* ; *Ceramic transactions*; American Ceramic Society: Westerville, Ohio, 2005; ISBN 1574981862.

90.   Panias, D.; Giannopoulou, I.P.; Perraki, T. Effect of synthesis parameters on the mechanical properties of fly ash-based geopolymers. *Colloids Surfaces A Physicochem. Eng. Asp.* 2007, *301*, 246–254. https://doi.org/10.1016/j.colsurfa.2006.12.064

91.   Dionisio, K.L.; Phillips, K.; Price, P.S.; Grulke, C.M.; Williams, A.; Biryol, D.; Hong, T.; Isaacs, K.K. Data Descriptor: The Chemical and Products Database, a resource for exposure-relevant data on chemicals in consumer products. *Sci. Data* 2018, *5*. https://doi.org/10.1038/sdata.2018.125

92.   Burduhos Nergis, D.D.; Vizureanu, P.; Corbu, O. Synthesis and characteristics of local fly ash based geopolymers mixed with natural aggregates. *Rev. Chim.* 2019, *70*.

93.   Kan, L. li; Lv, J. wei; Duan, B. bei; Wu, M. Self-healing of Engineered Geopolymer Composites prepared by fly ash and metakaolin. *Cem. Concr. Res.* 2019, *125*, 105895. https://doi.org/10.1016/j.cemconres.2019.105895

94.   Liu, X.; Ramos, M.J.; Nair, S.D.; Lee, H.; Nicolas Espinoza, D.; Van Oort, E. True self-healing geopolymer cements for improved zonal isolation and well abandonment. In Proceedings of the SPE/IADC Drilling Conference, Proceedings; Society of Petroleum Engineers (SPE), 2017; Vol. 2017-March, pp. 130–141.

95.   Zejak, R.; Nikolić, I.; Đurović, D.; Mugoša, B.P.; Blečić, D.; Radmilović, V. Influence of Na 2 O/Al 2 O 3 and SiO 2 /Al 2 O 3 ratios on the immobilization of Pb from electric arc furnace into the fly ash based geopolymers.. https://doi.org/10.1051/C

96.   Muñiz-Villarreal, M.S.; Manzano-Ramírez, A.; Sampieri-Bulbarela, S.; Gasca-Tirado, J.R.; Reyes-Araiza, J.L.; Rubio-Ávalos, J.C.; Pérez-Bueno, J.J.; Apatiga, L.M.; Zaldivar-Cadena, A.; Amigó-Borrás, V. The effect of temperature on the geopolymerization process of a metakaolin-based geopolymer. *Mater. Lett.* 2011, *65*, 995–998. https://doi.org/10.1016/j.matlet.2010.12.049

97.   Zhang, H.Y.; Kodur, V.; Wu, B.; Cao, L.; Wang, F. Thermal behavior and mechanical properties of geopolymer mortar after exposure to elevated temperatures. *Constr. Build. Mater.* 2016, *109*. https://doi.org/10.1016/j.conbuildmat.2016.01.043

98.   Al-Majidi, M.H.; Lampropoulos, A.; Cundy, A.; Meikle, S. Development of geopolymer mortar under ambient temperature for in situ applications. *Constr. Build. Mater.* 2016, *120*. https://doi.org/10.1016/j.conbuildmat.2016.05.085

99.   Fang, G.; Ho, W.K.; Tu, W.; Zhang, M. Workability and mechanical properties of alkali-activated fly ash-slag concrete cured at ambient temperature. *Constr. Build. Mater.* 2018, *172*. https://doi.org/10.1016/j.conbuildmat.2018.04.008

100.   Pangdaeng, S.; Phoo-ngernkham, T.; Sata, V.; Chindaprasirt, P. Influence of curing conditions on properties of high calcium fly ash geopolymer containing Portland cement as additive. *Mater. Des.* 2014, *53*, 269–274. https://doi.org/10.1016/j.matdes.2013.07.018

101.   Duxson, P.; Fernández-Jiménez, A.; Provis, J.L.; Lukey, G.C.; Palomo, A.; Van Deventer, J.S.J. Geopolymer technology: The current state of the art. *J. Mater. Sci.* 2007, *42*, 2917–2933. https://doi.org/10.1007/s10853-006-0637-z

102.   Hamidi, R.M.; Man, Z.; Azizli, K.A. Concentration of NaOH and the Effect on the Properties of Fly Ash Based Geopolymer. In Proceedings of the Procedia Engineering; Elsevier Ltd, 2016; Vol. 148, pp. 189–193.

103.   Abdul Rahi, R.H.; Azizli, K.A.; Man, Z.; Rahmiati, T.; Nuruddin, M.F. Effect of Sodium Hydroxide Concentration on the Mechanical Property of Non Sodium Silicate Fly Ash Based Geopolymer. *J. Appl. Sci.* 2014, *14*, 3381–3384. https://doi.org/10.3923/jas.2014.3381.3384

104.   Singh, B.; Ishwarya, G.; Gupta, M.; Bhattacharyya, S.K. Geopolymer concrete: A review of some recent developments. *Constr. Build. Mater.* 2015, *85*, 78–90.

105.   Yip, C.K.; Lukey, G.C.; van Deventer, J.S.J. The coexistence of geopolymeric gel and calcium silicate hydrate at the early stage of alkaline activation. *Cem. Concr. Res.* 2005, *35*, 1688–1697. https://doi.org/10.1016/j.cemconres.2004.10.042

106.   Hajimohammadi, A.; Provis, J.L.; Van Deventer, J.S.J. Effect of alumina release rate on the mechanism of geopolymer gel formation. *Chem. Mater.* 2010, *22*, 5199–5208. https://doi.org/10.1021/cm101151n

107.   Somna, K.; Jaturapitakkul, C.; Kajitvichyanukul, P.; Chindaprasirt, P. NaOH-activated ground fly ash geopolymer cured at ambient temperature. *Fuel* 2011, *90*, 2118–2124. https://doi.org/10.1016/j.fuel.2011.01.018

108.   Nazari, A., Sanjayan, G., Handbook of Low Carbon Concrete - Google books, available online, (accessed on Feb 16, 2020).

109.   Yahya, Z.; Husin, K.; Abdullah, M.M.A.B.; Ismail, K.N.; Razak, R.A. Strength, Density and Water Absoprtion of Palm Oil Boiler Ash (POBA) Geopolymer Brick/IBS Brick. *Key Eng. Mater.* 2016, *673*, 21–28.

110.   Ariffin, N.; Abdullah, M.M.A.B.; Mohd Arif Zainol, M.R.R.; Baltatu, M.S.; Jamaludin, L. Effect of Solid to Liquid Ratio on Heavy Metal Removal by Geopolymer-

Based Adsorbent. In Proceedings of the IOP Conference Series: Materials Science and Engineering; Institute of Physics Publishing, 2018; Vol. 374.

111.   Prasetya, F.A.; Sukmana, N.C.; Anggarini, U. Study of Solid-Liquid Ratio of Fly Ash Geopolymer as Water Absorbent Material. *MATEC Web Conf.* 2017, *97.*

112.   Chindaprasirt, P.; Chareerat, T.; Hatanaka, S.; Cao, T. High-Strength Geopolymer Using Fine High-Calcium Fly Ash. *J. Mater. Civ. Eng.* 2011, *23*, 264–270. https://doi.org/10.1061/(asce)mt.1943-5533.0000161

113.   Hardjito, D.; Wallah, S.E.; Sumajouw, D.M.J.; Rangan, B. V Fly Ash-Based Geopolymer Concrete. *Aust. J. Struct. Eng.* 2005, *6*, 77–86. https://doi.org/10.1080/13287982.2005.11464946

114.   Onprom, P.; Chaimoon, K.; Cheerarot, R. Influence of Bottom Ash Replacements as Fine Aggregate on the Property of Cellular Concrete with Various Foam Contents. *Adv. Mater. Sci. Eng.* 2015, *9*, 1–11. https://doi.org/10.1155/2015/381704

115.   Ghugal, Y.M. Effect of Fineness of Fly Ash on Flow and Compressive Strength of Geopolymer Concrete Article in Indian Concrete Journal. *Indian Concr. J.* 2013, *87*, 57–61.

116.   Palomo, A.; Krivenko, P.; Garcia-Lodeiro, I.; Kavalerova, E.; Maltseva, O.; Fernández-Jiménez, A. A review on alkaline activation: new analytical perspectives; Activación alcalina: Revisión y nuevas perspectivas de análisis. 2014, *64*, 22. https://doi.org/10.3989/mc.2014.00314

117.   Romisuhani, A.; AlBakri, M.M.; Kamarudin, H.; Andrei, S. V The Influence of Sintering Method on Kaolin-Based Geopolymer Ceramics with Addition of Ultra High Molecular Weight Polyethylene as Binder. *IOP Conf. Ser. Mater. Sci. Eng.* 2017, *267*, 12013. https://doi.org/10.1088/1757-899x/267/1/012013

118.   Rashad, A.M. A comprehensive overview about the influence of different admixtures and additives on the properties of alkali-activated fly ash. *Mater. Des.* 2014, *53*, 1005–1025. https://doi.org/10.1016/j.matdes.2013.07.074

119.   Ahmed, S.F.U. Fibre-reinforced geopolymer composites (FRGCs) for structural applications. In *Advances in Ceramic Matrix Composites*; Elsevier Ltd., 2014; pp. 471–495 ISBN 9780857091208.

120.   Nuaklong, P.; Sata, V.; Chindaprasirt, P. Influence of recycled aggregate on fly ash geopolymer concrete properties. *J. Clean. Prod.* 2016, *112*.

121.   Hadi, M.N.S.; Zhang, H.; Parkinson, S. Optimum mix design of geopolymer pastes and concretes cured in ambient condition based on compressive strength, setting time and workability. *J. Build. Eng.* 2019, *23*. https://doi.org/10.1016/j.jobe.2019.02.006

122.   Burduhos Nergis, D.D.; Vizureanu, P.; Andrusca, L.; Achitei, D.C. Performance of local fly ash geopolymers under different types of acids. *IOP Conf. Ser. Mater. Sci. Eng.* 2019, *572*, 012026. https://doi.org/10.1088/1757-899X/572/1/012026

123.   Komnitsas, K.; Zaharaki, D. Geopolymerisation: A review and prospects for the minerals industry. *Miner. Eng.* 2007, *20*, 1261–1277.

124.   Li, X.; Ma, X.; Zhang, S.; Zheng, E. Mechanical Properties and Microstructure of Class C Fly Ash-Based Geopolymer Paste and Mortar. *Materials (Basel).* 2013, *6*, 1485–1495. https://doi.org/10.3390/ma6041485

125.   https://www.holcim.ro/sites/romania/files/documents/Manual_de_utilizare_a_betoanelor_5.pdf (accessed on Feb 17, 2020).

126.   Kaja, A.M.; Lazaro, A.; Yu, Q.L. Effects of Portland cement on activation mechanism of class F fly ash geopolymer cured under ambient conditions. *Constr. Build. Mater.* 2018, *189*. https://doi.org/10.1016/j.conbuildmat.2018.09.065

127.   Chindaprasirt, P.; Phoo-ngernkham, T.; Hanjitsuwan, S.; Horpibulsuk, S.; Poowancum, A.; Injorhor, B. Effect of calcium-rich compounds on setting time and strength development of alkali-activated fly ash cured at ambient temperature. *Case Stud. Constr. Mater.* 2018, *9*. https://doi.org/10.1016/j.cscm.2018.e00198

128.   Azevedo, A.G.S.; Strecker, K.; Barros, L.A.; Tonholo, L.F.; Lombardi, C.T. Effect of curing temperature, activator solution composition and particle size in Brazilian fly-ash based geopolymer production. *Mater. Res.* 2019, *22*. https://doi.org/10.1590/1980-5373-MR-2018-0842

129.   Heah, C.Y.; Kamarudin, H.; Mustafa, A.M.; Bakri, A.; Binhussain, M.; Luqman, M.; Khairul Nizar, I.; Ruzaidi, C.M.; Liew, Y.M. Effect of Curing Profile on Kaolin-based Geopolymers. *Phys. Sci. Technol.* 2011, *22*, 305–311. https://doi.org/10.1016/j.phpro.2011.11.048

130.   Luna-Galiano, Y.; Fernández-Pereira, C.; Izquierdo, M. Contributions to the study of porosity in fly ash-based geopolymers. Relationship between degree of reaction, porosity and compressive strength. *Mater. Constr.* 2016, *66*, e098. https://doi.org/10.3989/mc.2016.10215

131.   Amar S., B.; Srinivas, N. Effect of curing temperature on compressive strength of geopolymer concrete. *Int. J. Recent Sci. Res.* 2019, *7*, 12377–12381. https://doi.org/10.24327/IJRSR

132.   Mo, B.H.; Zhu, H.; Cui, X.M.; He, Y.; Gong, S.Y. Effect of curing temperature on geopolymerization of metakaolin-based geopolymers. *Appl. Clay Sci.* 2014, *99*, 144–148. https://doi.org/10.1016/j.clay.2014.06.024

133.   Palomo, A.; Grutzeck, M.W.; Blanco, M.T. Alkali-activated fly ashes: A cement for the future. *Cem. Concr. Res.* 1999, *29*, 1323–1329. https://doi.org/10.1016/S0008-8846(98)00243-9

134.   Rovnaník, P. Effect of curing temperature on the development of hard structure of metakaolin-based geopolymer. *Constr. Build. Mater.* 2010, *24*, 1176–1183. https://doi.org/10.1016/j.conbuildmat.2009.12.023

135.   Zhang, R.; Ma, D. Effects of Curing Time on the Mechanical Property and Microstructure Characteristics of Metakaolin-Based Geopolymer Cement-Stabilized Silty Clay. *Adv. Mater. Sci. Eng.* 2020, 9605941. https://doi.org/10.1155/2020/9605941

136.   Mizerová, C.; Kusák, I.; Rovnaník, P. Electrical properties of fly ash geopolymer composites with graphite conductive admixtures. *Acta Polytech. CTU Proc.* 2019, *22*, 72–76. https://doi.org/10.14311/app.2019.22.0072

137.   Temuujin, J.; Minjigmaa, A.; Lee, M.; Chen-Tan, N.; van Riessen, A. Characterisation of class F fly ash geopolymer pastes immersed in acid and alkaline solutions. *Cem. Concr. Compos.* 2011, *33*, 1086–1091. https://doi.org/10.1016/j.cemconcomp.2011.08.008

138.   Hardjito, D.; Cheak, C.C.; Ing, C.H.L. Strength and Setting Times of Low Calcium Fly Ash-based Geopolymer Mortar. *Mod. Appl. Sci.* 2008, *2*. https://doi.org/10.5539/mas.v2n4p3

139.   Ariffin, M.A.M.; Bhutta, M.A.R.; Hussin, M.W.; Tahir, M.M.; Aziah, N. Sulfuric acid resistance of blended ash geopolymer concrete. *Constr. Build. Mater.* 2013, *43*, 80–86. https://doi.org/10.1016/j.conbuildmat.2013.01.018

140.   Rajamane, N.P.; Ambily, P.S. Modified Bolomey equation for strengths of lightweight concretes containing fly ash aggregates. *Mag. Concr. Res.* 2012, *64*, 285–293. https://doi.org/10.1680/macr.11.00157

141.   Ridtirud, C.; Chindaprasirt, P.; Pimraksa, K. Factors affecting the shrinkage of fly ash geopolymers. *Int. J. Miner. Metall. Mater.* 2011, *18*, 100–104. https://doi.org/10.1007/s12613-011-0407-z

142.   Degirmenci, F.N. Freeze-Thaw and fire resistance of geopolymer mortar based on natural and waste pozzolans. *Ceram. - Silikaty* 2018, *62*, 41–49. https://doi.org/10.13168/cs.2017.0043

143.   Pilehvar, S.; Cao, V.D.; Szczotok, A.M.; Valentini, L.; Salvioni, D.; Magistri, M.; Pamies, R.; Kjøniksen, A.-L. Mechanical properties and microscale changes of geopolymer concrete and Portland cement concrete containing micro-encapsulated phase change materials. *Cem. Concr. Res.* 2017, *100*, 341–349. https://doi.org/10.1016/j.cemconres.2017.07.012

144.   Bernal, S.A.; Provis, J.L.; Walkley, B.; San Nicolas, R.; Gehman, J.D.; Brice, D.G.; Kilcullen, A.R.; Duxson, P.; Van Deventer, J.S.J. Gel nanostructure in alkali-activated binders based on slag and fly ash, and effects of accelerated carbonation. *Cem. Concr. Res.* 2013, *53*, 127–144. https://doi.org/10.1016/j.cemconres.2013.06.007

145.   Contribuții privind obținerea de geopolimeri prin valorificarea unor reziduuri de producție. *Galati Univ. Press - Ed. Univ. "Dunărea Jos" din Galați, 2016.*

146.   Olivia, M.; Nikraz, H. Properties of fly ash geopolymer concrete designed by Taguchi method. *Mater. Des.* 2012, *36*, 191–198. https://doi.org/10.1016/j.matdes.2011.10.036

147.   Bondar, D.; Lynsdale, C.J.; Milestone, N.B.; Hassani, N.; Ramezanianpour, A.A. Effect of adding mineral additives to alkali-activated natural pozzolan paste. *Constr. Build. Mater.* 2011, *25*, 2906–2910. https://doi.org/10.1016/j.conbuildmat.2010.12.031

3. General Objectives and Methodology of Experimental Research

3.1. Research objectives

The book presents a complex and interdisciplinary study in the field of physics, chemistry, materials science and civil engineering related to the designing and elaboration of new oxide materials based on mineral wastes through the geopolymerization chemical reaction.

The innovations of the proposed approach are (i) reducing the carbon footprint and energy consumption by creating geopolymers at relatively low temperatures compared to present conventional materials, (ii) two types of waste - power plant ash and glass powder - and (iii) obtaining improved or comparable properties to Portland cement-based materials without involving Portland cement in the mixture.

The book topic is related with fundamental and applied research in the field of materials engineering, because it aims to present the obtaining and characterization of materials with original chemical compositions that can be used in the field of construction of building facades, floors, decorative panels, insulating bricks, furniture garden etc. Furthermore, these materials will be obtained at low temperatures (<100 °C), with minimal energy consumption for the curing process, while for bricks or other types of clay products a high temperature oxidation step is required (> 800 °C) [1,2].

In order to fulfill the main research objectives, the following scientific and technical objectives are taken into account:

• Formulation of the technological concept by designing the method of obtaining new materials based on thermal power plant ash, glass powder and sand. Different types of matrices and percentages of reinforcing particles will be taken into account, respectively scientific research methods involving appropriate equipment. This stage involves (i) the design of the methodology for obtaining geopolymers, (ii) the establishment of materials investigation procedures and (iii) the collection of mineral waste and the purchase of raw materials.

• Demonstration of the concept by (i) characterization of raw materials (thermal power plant ash, glass powder and sand) by physical-chemical and morpho-structural analysis, (ii) development of a new geopolymer based on thermal power plant ash activated by the chemical geopolymerization reaction and (iii) strengthening the geopolymer with glass powder and sand in order to improve the mechanical properties.

• Validation of technology (results), by characterizing the materials obtained, from a chemical, structural, physical-mechanical and thermal point of view. This step involves (i) chemical characterization by electron microscopy and X-ray fluorescence spectroscopy, (ii) structural characterization by light and electron microscopy, (iii) mechanical characterization of the geopolymer obtained in terms of compressive strength; and bending strength and (iv) evaluation of the thermal behavior by determining the evolution of the sample mass during heating simultaneously with the study of phase transformations by differential thermal analysis.

The presented research objectives can be achieved following the observance of the experimental plan (Table 3.1), consisting of 4 stages, which include the theoretical study, design, development/obtaining and characterization of geopolymers.

Table 3.1. Research methodology. Experimental plan stages.

Geopolymers designing	Establishing the stages of ash preparation
	Establishing the stages of reinforcing particles preparation
	Establishing the alkaline activator
	Establishing the solid to liquid ratio
	Establishing the curing process parameters
Obtaining of geopolymers	Drying and sifting of raw materials
	Mixing the power plant ash with a activator solution based on sodium silicate and sodium hydroxide
	Introduction of reinforcing particles into the mixture
	Pouring into molds and drying the mixture
	Obtaining the geopolymers (samples)
Geopolymers characterization	Oxide chemical composition investigation by XRF and EDS
	Microstructural characterization employing optical and electron microscopy
	Mineralogical characterization by XRD and FTIR
	Setting time evaluation by Vicat method
	Relative pore size distribution evaluation by RMN
	Compressive strength evaluation
	Flexural strength evaluation
	High temperatures stability analysis
	Phase transformation evaluation
Results and discussions	Critical analysis and correlation of the experimental results

Obtaining and characterizing geopolymer samples based on thermal power plant ash involves the use of several types of equipment that provide information on chemical, structural, physical-mechanical and durability characteristics following laboratory investigations, performed on specimens specific to each test.

3.2. Experimental methodology

In order to design and obtain a geopolymer based on thermal power plant ash of interest on the consumer goods market, a number of characteristics must be taken into account, such as chemical, structural, mechanical (high resistance to compression and bending, etc.) and thermal (thermal conductivity, phase transformations during heating etc.). The identification of possible applications for the obtained geopolymer will be made following its characterization in laboratory conditions according to the properties mentioned above.

In Table 3.2. are listed the methods of characterization were performed on the coal-ash based geopolymers obtained.

Table 3.2. Characterization methods applied on the obtained geopolymers.

Laboratory investigations	Characterization method
Chemical characterization	Energy-dispersive X-ray spectroscopy (EDS)
	X-ray fluorescence spectroscopy (XRF)
Structural characterization	Optical microscopy (OM)
	Scanning Electron Microscopy (SEM)
	Fourier Transform Infrared Spectroscopy (FTIR)
	X-ray diffraction (XRD)
Physical-mechanical characterization	Setting time (Vicat)
	Relative pore size distribution (NMR)
	Compressive strength test
	Flexural strength test
Thermal behavior characterization	Thermogravimetric analysis (TGA)
	Differential thermal analysis (DTA)

The analysis of the chemical composition, mainly of the percentages of Al, Si and Ca, is performed in order to correctly establish the solution mixture necessary for the activation of the geopolymers based on local ash from the power plant. The main purpose of the structural characterization is to highlight the bonds created between the particles, the degree of dissolution of the ash, as well as the distribution of the reinforcing elements in the sample volume.

Obtaining and characterizing geopolymer samples based on local ash from power plants involves the use of several types of equipment that provide information on chemical, structural, physical-mechanical and thermal characteristics following laboratory investigations, performed on specimens specific to each test.

The identification of the engineering field with the highest potential for the use of the obtained materials will be established according to the properties they will possess.

3.3. Analyzing methods and equipment's used in the experimental plan

In addition to the type of alkaline activator and the parameters of the curing process, the characteristics of the raw material (chemical composition, moisture, particle size etc.) play an essential role in developing the mechanical strength properties, durability and microstructure of the resulting material. In this research, the geopolymers were analyzed by specific methods of chemical, structural, physical-mechanical and thermal characterization.

In order to evaluate and identify the class of the obtained materials, depending on the acquired properties, the results of the tests performed were compared with those in the literature or the standards in force. The synthetic description of the equipment used for each determination are presented further.

3.3.1 Samples and raw materials drying

ED series natural convection ovens are used for sterilizing/drying laboratory samples and performing analyzes at a given temperature for microbiological samples; biological and especially for those in powder form. In the case of the binder type stope, the temperature is regulated by an electronic controller over a maximum time range of 99 hours. The temperature can be set step by step and on a single ramp.

This equipment can be used in the pharmaceutical and chemical industries, biotechnology, medicine, veterinary medicine, universities, research institutes, the food industry, drying and sterilization applications that do not require high drying speeds or special treatment programs are also ideal for keeping certain products warm [3].

This type of stove has an internal/useful volume of 116 lt. and multiple shelfs, therefore it can be used to dry or cure multiple samples at the same time. Also, it operates in a temperature range of 5 to 300 °C, with a temperature variation at 150 °C of $1.7 \pm$ K and a temperature fluctuation at 150 °C of $0.3 \pm$ K. Moreover, in the case when it is necessary to open the door, the returning time at 150 °C is of 5 min. The necessary time for heating up to 150 °C is 19 min.

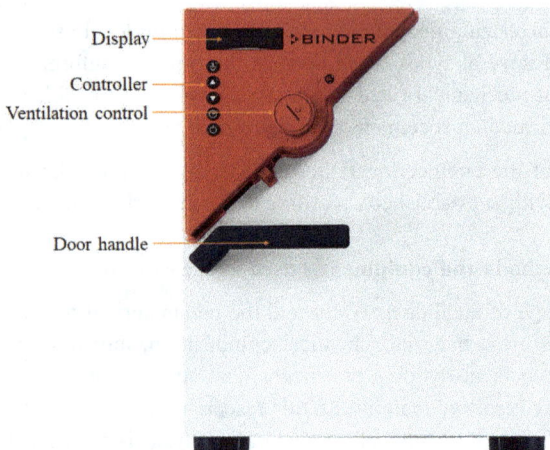

Figure 3.1. The interface of the Binder type stove used for samples drying [3].

3.3.2   Particle size separation

All the raw materials have been sifted using a *Vibratory Sieve Shaker AS 200 (Figure 3.2)* with the following parameters: sifting time - 20 min; amplitude - 2 mm; interval operation: 20 sec.

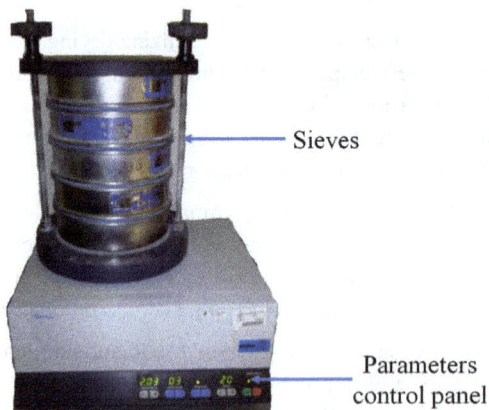

Figure 3.2. Vibratory Sieve Shaker AS 200 control [4].

The analytical sieve shakers of the series AS 200 are used in research & development, quality control of raw materials, interim and finished products as well as in production monitoring. The controllable electromagnetic drive offers an optimal adaption for every product. Sharp fractions are obtained even after very short sieving times.

Table 3.1. Vibratory Sieve Shaker AS 200 control features [5].

Applications	separation, fractioning, particle size determination
Field of application	agriculture, biology, chemistry / plastics, construction materials, engineering / electronics, environment / recycling, food, geology / metallurgy, glass / ceramics, medicine / pharmaceuticals
Feed material	powders, bulk materials, suspensions
Measuring range*	20 μm - 25 mm
Sieving motion	throwing motion with angular momentum
Max. batch / feed capacity	3 kg
Max. number of fractions	9 / 17
Max. mass of sieve stack	6 kg
Amplitude	digital, 0.2 - 3 mm
Sieve acceleration	1.0 - 15.1 g
Time display	digital, 1 - 99 min
Interval operation	10 - 99 s
Storable SOPs	9
Suitable for dry sieving	yes
Suitable for wet sieving	yes
Serial interface	yes
Including test certificate / can be calibrated	yes
Suitable sieve diameters	100 mm / 150 mm / 200 mm / 203 mm (8")
Max. height of sieve stack	450 mm
Clamping devices	"standard", "comfort", each for wet and dry sieving
Standards	CE

All sieve shakers of the series AS 200 work with an electromagnetic drive that is patented by RETSCH (EP 0642844). This drive produces a 3D throwing motion that moves the product to be sieved equally over the whole sieving surface. The advantage: high stress capacity, extremely smooth operation, and short sieving times with high separation efficiency.

3.3.3   *Quantitative measurements*

The ADAM PW 254 scales (Figure 3.3) are ideal for laboratory and routine weighing. The scale can also be used for some advanced weighing functions, such as components measuring for drugs or explosives.

Adam Equipment analytical balances are equipped with a backlit LCD display that is easy to read and at the same time displays the weight and results for a specific application. The stainless steel weighing plate is corrosion resistant and easy to clean. The balance has an internal calibration system that provides accurate measurements based on user-defined time and temperature limits, and a container tracker for visual inspection to prevent overloading of the balance. This scale also supports external calibration. An RS-232 port on the rear panel is available for easy connection to a printer or computer, while integrated features allow GLP (Good Laboratory Practices) to be traced. The weighing hook under the balance suspends the samples under the balance for measuring weighing and density [6].

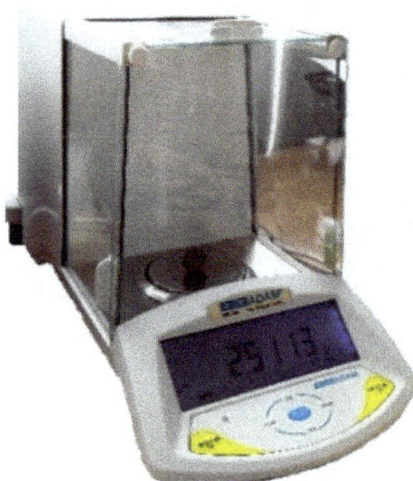

Figure 3.3. ADAM PW 254 balance [6].

When weighing the samples, the balance may be set to show the weight above or below the upper or lower limit. The arrows below the bar graph will appear on the display to indicate that weighing is in progress. Arrows and bars between arrows indicate when the balance is below the lower limit, between the upper limit or the upper limit.

The vibrator can be set to operate when the scale is out of range (below the lower or upper limit), on (above the lower level and below the upper limit), or off. Just set a limit if you want. If only one limit is set, the remaining limit is considered zero (low) or high (high). Check that the scale does not work if the weight is less than 20 days. This is the minimum weight at which the indicator bars are displayed, and the ringtone will ring if it is not turned off [7].

The covering box is metallic and must be supported on all supports (to dampen the vibrations it would be advisable to sit on a bed of sand). This equipment uses the electromagnetic balance system with a built-in microprocessor that controls the measurement function, data processing, the calculation being a electronic reading balance (computerized balance) with special effect in reducing errors speed and reduction of errors in weighing and data processing.

Installation and operating instructions:

- Choose a suitable place to install the balance:
    - away from heat corrosive substances, vibrations, drafts;
    - place the tray on the support;
    - the current connection is made by operating the power switch;
    - after connecting to the power, wait 30 minutes for heating;
    - balance the balance with the help of the balancing screws so that the air bubble is placed in the center of the circle of the level indicator.
- An automatic test of the display screen is performed, resulting in the appearance of the digits 0.00 on the screen:
    - after the display test, the message OFF is displayed.
- Press ON / OFF (ON-active mode):
    - then calibrate the balance using the standard weight of (2 kg) 2000gr., press CAL;
    - when the balance is empty (the balance is calibrated whenever the balance is moved from the original position; the calibration is checked periodically; do not throw heavy objects on the balance plate;
    - after pressing CAL flashes on the screen with flashing light the value of the standard calibration weight (2000 gr.) appears, after the flashing ends, the

screen lights up steadily indicating the value of the calibration weight (standard) in grams -2000;

  o expect the beep sound after which the weight on the plate can be lifted when the balance is ready for weighing;

  o when the ERROR screen appears, the above operations are repeated.

- When the material is not weighed directly on the plate and a vessel (container = tray) is used, it is calibrated as follows:

  o tapping the TA RE key allows the display to be reset to zero whenever desired, either for a tray, a container that can be placed on the turntable (whose weight does not exceed the capacity of the balance) and to correct the display whenever it does not appear zero;

  o the container is placed on the tray;

  o press the TA RE key - a series of intermittent horizontal lines appears on the screen until stability is reached, after which the value 0.00 will appear on the screen.

- When the balance is engaged the balance (ON) weighing is done in grams (gr.) To change the unit of measurement:

- Press the MENU key successively until the desired unit of measure appears;

  o after this display press for confirmation of the ENTER key by actuating this unit of measure;

  o to return to grams, press the MENU key successively.

Maintenance:

- not throwing heavy objects on the turntable;
- not spray (wet) with water directly on the scale;
- clean the banana with a soft cloth dampened with detergent water (not acetone or other solvents for cleaning).

3.3.4   Components mixing

The planetary mixer (Figure 3.4) is used to mix polygranular sand with cement, and comes with a paddle that is ideal for mixing powdery materials. Technical specifications: mass 38 kg, container capacity 5 liters, power 120 W, planetary rotation speed max. 125 rpm, min. 62 rpm, blade rotation speed, max 285 rpm, min 140 rpm.

Install the desired stirrer / vane by inserting, lifting and turning it clockwise. Use lever 5 to lift the tub, close the protective grille clockwise until it locks.

When entering the working time, keep in mind that the indicated number is only a reference value and does not correspond to the working minutes.

Attention: when the protection grille is raised, the motor stops, but the timer does not return to zero but the counting continues. It is therefore recommended that you wait for the set time to elapse. After any mixing process, reset the timer.

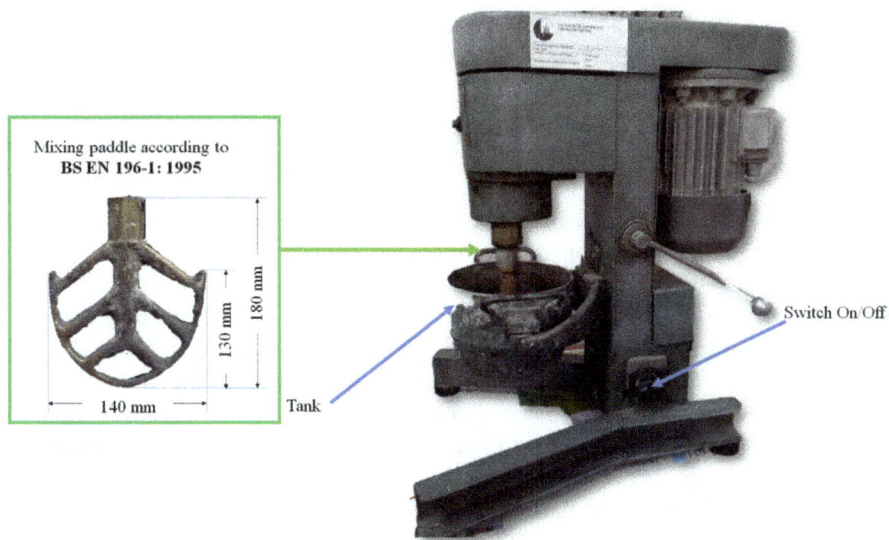

Figure 3.4. Variable speed mixer [8].

It is recommended to use a plastic spatula to clean the appliance and remove the residue that remains in the bowl. The protective grille, subjected to spraying, is made of stainless steel and can be washed, as well as the tub, with water and detergent. The appliance is enameled and can be cleaned with a damp sponge.

3.3.5   Chemical and structural characterization methods

In order to identify materials with geopolymerization potential, it is necessary to evaluate their chemical composition. The quality of the final structure and properties results mainly from the uniformity and strength of the tetrahedral structure formed by the Si-O-Al system. Any impurity in the base material, especially the calcium content, can lead to the formation of defects in the structure that significantly influence the final properties of the material. According to the experimental plan, the determination of the elemental and oxide chemical composition is the most important step in order to identify the geopolymerization potential of the raw material and to establish the correct activation solution.

3.3.6   *Energy-dispersive X-ray spectroscopy*

Energy Dispersive Spectroscopy (EDS) is used to qualitatively determine the chemical elements in a sample and to estimate their distribution. The accuracy of the quantitative determination of chemical elements may be affected by various factors related to the type of sample and its constituents. Due to the simplicity and speed of the method, this type of analysis is ideal for identifying the geopolymerization potential of the raw material [9].

To stimulate the emission of characteristic X-rays from a sample, a beam of high-energy charged particles, electrons, protons or an X-ray beam, is focused on the studied sample. Normally, an atom in the sample contains unexcited electrons that rotate in orbits around the nucleus. The incident beam can excite an electron from an outer layer, thus removing it from orbit, leaving a vacant position in its place. This vacancy is then occupied by an electron with a higher energy that releases the energy difference in the form of X-rays. The number and energy of the emitted samples of the sample can be measured using an energy dispersion spectrometer and compared with a database. in order to identify the chemical elements present. Figure 3.5 shows schematically the physical mechanism of expulsion of the electron under the action of the incident beam and the occupation of the vacant position by an electron with higher energy from an outer layer.

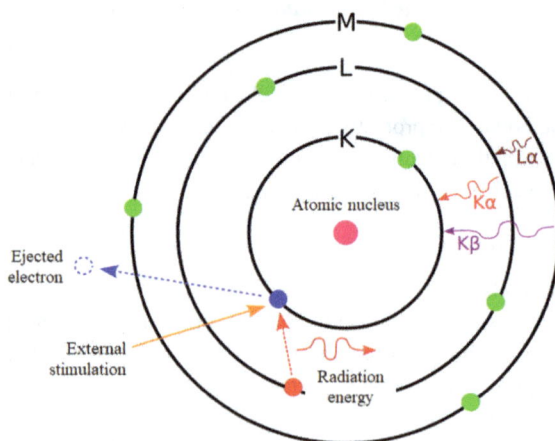

Figure 3.5. Schematic representation of the specific physical mechanism of EDS [9].

Most of the materials with high content of silicon and aluminum oxides can be used in the manufacture of geopolymers. During their discovery and development, natural minerals were mainly used, because they were easy to identify and exploit, taking into account the fact that approximately 65% of the Earth's crust is made of silicon and aluminum-based materials.

For the qualitative determination of the chemical composition specific to geopolymers, the SEM VegaTescan II equipment was used, which consists of a Bruker type EDS detector with the help of which information related to the chemical composition of the analyzed sample is taken. The determinations were performed on a minimum of five samples (quantities of thermal power powder) taken from different areas of the thermal power plant ash and on a minimum of three samples of glass powder. Prior to the examination, the powders were subjected to a drying step in order to eliminate the influence of water on the analysis.

3.3.7   X-ray fluorescence spectroscopy

In order to obtain environmentally friendly materials and to conserve natural resources, the possibility of obtaining the geopolymers from various industrial wastes, such as thermal power plant ash and glass powder, was studied. This waste is inevitably heterogeneous, and certain impurities, such as calcium oxides and iron oxides in the basic components, lead to the creation of side effects during the geopolymerization reaction. This can lead to changes in curing times or mechanical properties specific to the final product. Accurate qualitative determination of the chemical elements in the raw material used is essential for the correct activation of the geopolymer.

X-ray Fluorescence Spectroscopy (XRF) is based on the emission of fluorescent X-rays from a material that has been excited by high-energy or gamma-ray bombardment (Figure 3.6). When samples are exposed to short-wavelength or gamma-ray X-rays, ionization of the atoms in the composition occurs. Ionization consists of the expulsion of one or more electrons from an atom and can occur if the atom is exposed to radiation with an energy higher than its ionization energy. The elimination of an electron in this way leads to the destabilization of the electronic structure of the atom, so the electrons in the upper orbits pass into lower orbits to fill the vacancies left behind. At this transition from the upper or lower orbits, the electrons release energy in the form of photons, whose energy is equal to the energy difference between the two orbits involved. Thus, the material emits radiation with energy characteristic of the atoms present. The term fluorescence applies to phenomena in which the absorption of radiation from one energy results in the re-emission of radiation from another energy (generally lower) [10].

Figure 3.6. Schematic representation of the specific physical mechanism XRF [11].

The determination of the quantitative chemical composition of the samples was performed using an equipment S8 Tiger Bruker (Figure 3.7) which consists of a sequential spectrometer, a goniometer with 4-position collimator exchanger and two scintillation and gas detectors that are equipped Center for Eco-Metallurgical Research and Expertise (ECOMET UPB). The X-ray tube of its composition is provided with a demineralized water cooling system, and the vacuum system is based on a rotary pump.

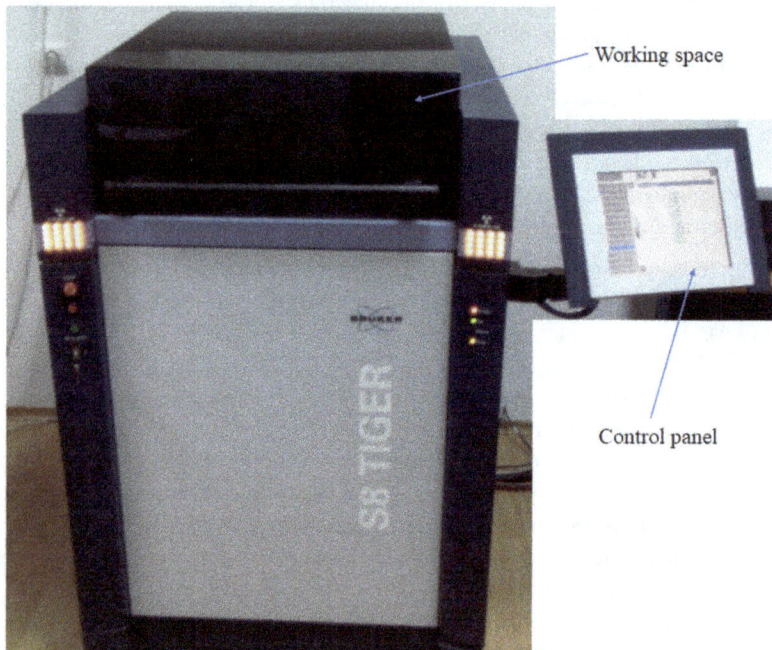

Figure 3.7. X-ray fluorescence spectrometer type S8 TIGER [12].

X-ray Fluorescence Spectroscopy (XRF) is generally used to determine the elemental or oxide chemical composition, especially of glass, ceramic materials or building materials. The determination of the quantitative chemical composition of the samples was performed with by means of the Bruker Spectrometer equipment type Tiger S8 (Bruker, Berlin, Germany).

3.3.8   *Microstructural analysis by optical microscopy*

In the case of geopolymers, the microstructure is based on small crystals called grains. In general, the smaller the grain size, the better the material obtained in terms of mechanical properties and denser. By means of the optical microscope, the grains formed as a result of the geopolymerization reaction between the ash of the thermal power plant and the alkaline activator are highlighted, as well as the shape, distribution or interface between the glass, sand or gravel particles and matrix.

The microstructure of geopolymers plays an important role in acquiring their properties, and this can be analyzed using an optical microscope that consists of a light source that projects a light beam on the surface of the sample. It is reflected by the analyzed surface, passes through a lens and forms an inverted image. The image is then inverted and magnified by the eyepiece (Figure 3.8), and can be seen with the naked eye or with digital devices.

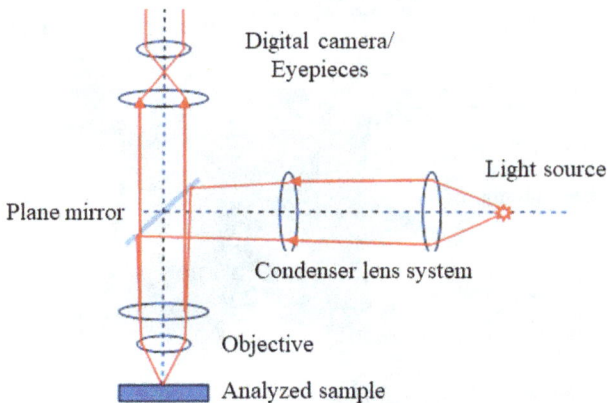

Figure 3.8. The principle of operation of an optical microscope.

In the case of geopolymers, the microstructure is made up of small crystals called grains. In general, the smaller the grain size, the better the material obtained in terms of mechanical properties and denser. With the help of the optical microscope, the grains formed as a result of the geopolymerization reaction between the power plant ash and the alkaline activator are highlighted, as well as the shape, distribution or interface between the glass particles, sand or gravel and matrix.

The Zeiss optical microscope (Figure 3.9) from the endowment of the Laboratory for scientific investigation and conservation of cultural heritage, within the Arheoinvest Interdisciplinary Platform, within the "Alexandru Ioan Cuza" University of Iasi [8] is composed of three main systems:

- lighting system consisting of filters (BF and DF), eyepieces, lenses and a series of prisms and mirrors.
- digital image capture and transmission system.
- mechanical system for adjusting the position of the samples.

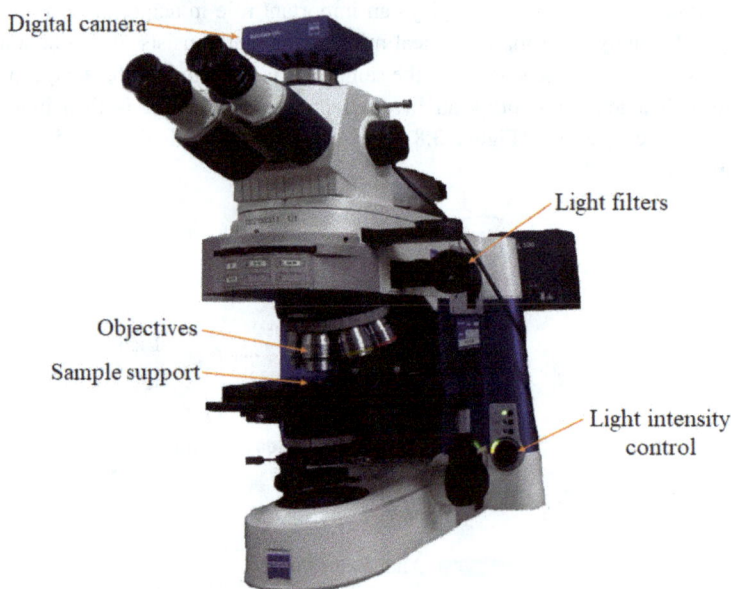

Figure 3.9. Optical microscope type ZEISS [13].

Morphological analysis of the samples reveals the formation of a heterogeneous matrix consisting of a vitrified gel with a high density of pores distributed throughout the volume. Also, with the help of the optical microscope, important information related to the mechanism of advancement of cracks during mechanical stresses can be obtained. This phenomenon can be observed only on the tested samples, unprepared in terms of the flatness of the analyzed surface. In order to study the distribution of pores and compounds (aggregates, crystals, etc.) the surface of the samples must be prepared by sanding and blowing with air.

While studying the cracking mechanism, chips resulting from the destruction of the samples during compression or bending strength tests were used. For the evaluations of other characteristics, such as the connection between the components, the distribution and geometry of the pores, the homogeneity of the mixture, $10 \times 10 \times 10$ mm^3 samples obtained by cutting $20 \times 10 \times 10$ mm^3 samples were used (Figure 3.10.).

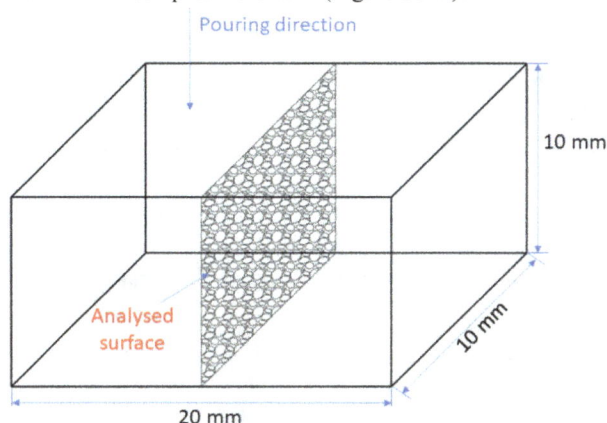

Figure 3.10. Sample preparation for the surface analysis.

Figure 3.11. Grinding and polishing machine Forcipol 2V [14].

In order to obtain samples with plane surfaces, necessary to study the structure of geopolymers employing an optical microscope, the FORCIPOL 2V grinding machine (Figure 3.11.) was used. The equipment is equipped with two discs with a diameter of 250 mm whose maximum speed can reach 600 rpm. and can be controlled with a precision of ± 1 rpm. The abrasive paper with the desired granulation is mounted on the surface of the discs, depending on the roughness of the surface of the analyzed sample, then the sample is polished until a smooth surface is obtained.

3.3.9   Microstructural analysis by scanning electron microscopy

Through electron microscopy, information on the morphology of the analyzed surfaces can be obtained. Therefore, it highlights the connection between the components (matrix and reinforcing particles), the distribution and geometry of the pores, the homogeneity of the mixture, the cracking/destruction mechanism, in the case of samples tested in terms of mechanical properties, the degree of dissolution of the component. base etc.

A scanning electron microscope (SEM) contains an electron source that produces a high-energy electron beam. By means of an electromagnetic system the beam is focused to a diameter of about 100 Å. When the electrons in the beam hit the atoms in the analyzed sample, they absorb energy and release their own electrons, known as secondary electrons, which are captured by a detector (Figure 3.12). After analyzing the captured electrons, the software forms a representative image of the surface.

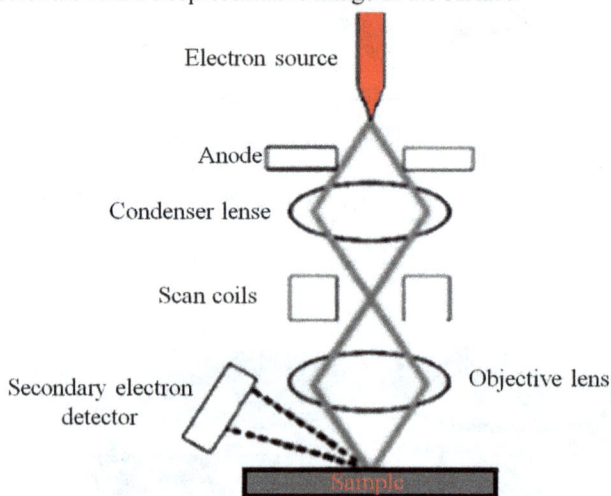

Figure 3.12. Schematic representation of the component elements in an SEM [15,16].

The scanning electron microscope type FEI Quanta FEG 450 used to study the microstructure of geopolymers has an amplification power of 100 kX and is equipped with a secondary electron detector (SE) with which images can be obtained at high amplification rates (5000 ÷ 10000X) compared to the optical microscope. Moreover, by means of the secondary electron detector (EDS) the chemical composition or the distribution of the chemical elements in the studied sample can be determined.

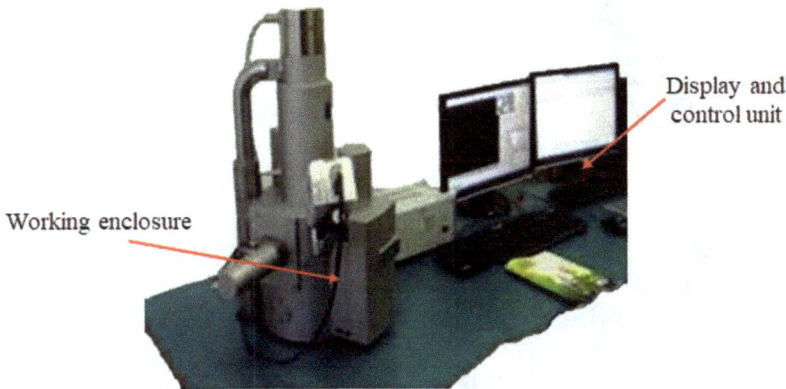

Display and control unit

Working enclosure

Figure 3.13. Scanning electron microscope type SEM Quanta 200 3D – FEI Signal detector type Apollo X SDD [17].

Unlike transmission electron microscopy (TEM) where the emerging electron beam contains the entire image of the analyzed specimen, in the case of SEM, at a certain point in time, the emerging beam may contain only local information (a 'pixel') in the image. In order to reproduce the entire image, the electron beam needs to be scanned over the entire surface of the specimen [18].

Due to the very low electrical conductivity of the obtained geopolymeric materials, in order to obtain a representative image by means of the scanning electron microscope, it was necessary to cover their surface with a layer of graphite. The main purpose of the coating is to increase the electrical conductivity of the samples and decrease the background noise of the analysis. The equipment (Figure 3.14) used for depositing the graphite layer consists of a vacuum enclosure in which the samples are positioned under a graphite cord mounted between two terminals, which are part of a circuit, positioned on the enclosure cover. When the current passes through the circuit, the graphite cord is heated to luminescence, emitting carbon atoms that deposit on the surface of the samples inside. In order to obtain a uniform coating, the enclosure is emptied with a vacuum pump.

Also, the geopolymer sample could be analysed in environmental scanning electron microscopy (ESEM). Its importance derives from the fact that it is the only type of electron microscope that is not limited to vacuum operation. For this reason, it can be used to examine living cells in their natural environment or in atmospheric air.

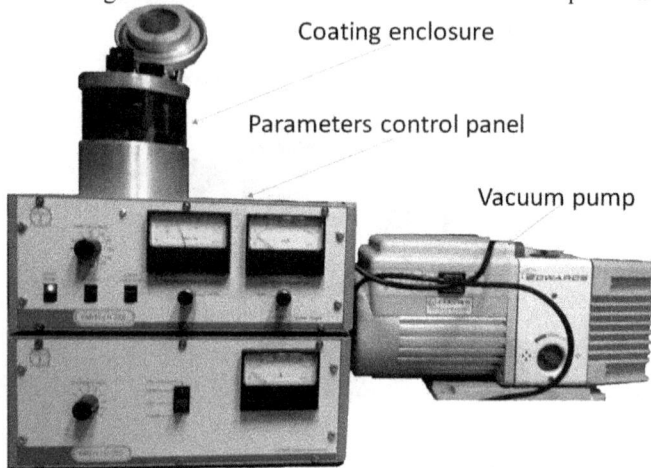

Coating enclosure

Parameters control panel

Vacuum pump

Figure 3.14. Laboratory equipment for graphite coatings [19].

The great innovation of this device consists in the differential pumping system: between the electron source (which has been in advanced vacuum) and the specimen (at high pressure) there are several different levels of pumping. Although the electron beam suffers more and more dispersions as the high pressure level of the specimen chamber approaches, if a sufficiently good ratio is maintained between the number of undistorted and deviated electrons, the noise produced by the latter maybe negligible. Apart from these differences, in the case of ESEM the image is formed as in the case of SEM [18].

3.3.10 Mineralogical analysis by means of X-Ray diffraction

X-Ray Diffraction (XRD) is a non-destructive technique used to identify and quantitatively evaluate the crystalline phases of different materials.

The operation principle is based on the X-ray bombardment (Figure 3.15), obtained in a X-ray tube, of the analyzed sample, which is introduced into a vacuum chamber. This chamber contains a copper anode and an incandescent cathode that emits electrons.

For the structural X-ray analysis of the obtained geopolymers, a PANalytical X'Pert PRO MRD X-ray Diffractometer (Figure 3.16) was used.

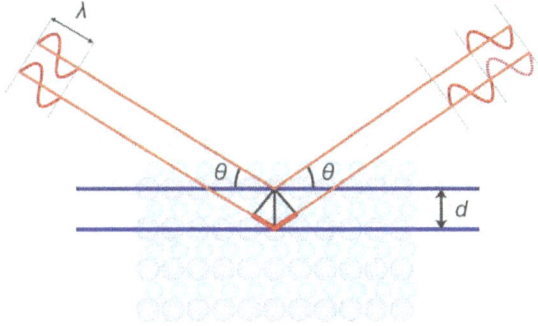

Figure 3.15. Schematic representation of the specific physical mechanism XRD [20].

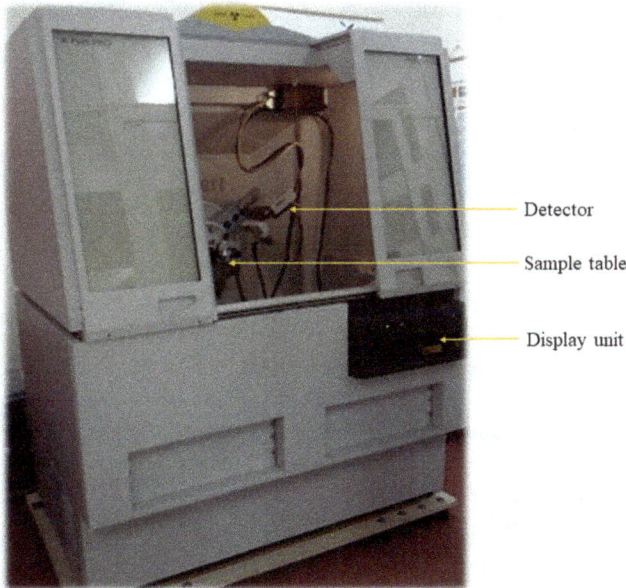

Figure 3.16. X-ray diffractometer type X'Pert Pro MPD [21].

The determination of the phases from the analyzed samples was performed on powders obtained by grinding the geopolymer samples with the dimensions of 20x20x25 mm^3.

3.3.11 Mineralogical characterization by Fourier transform infrared spectroscopy

Fourier transform infrared spectroscopy (FTIR) is a non-destructive technique for analyzing the chemical structure of a material that consists of obtaining an infrared (IR) spectrum of light absorption at different wavelengths. a bundle. The absorption of IR light is caused by the change in the state of vibration between the atoms of the molecules present on the surface of the analyzed organic or inorganic sample.

In the case of a molecule composed of three atoms (for example the water molecule) IR light can produce tensile and compressive vibrations (Figure 3.17.a) of the bonds between hydrogen and oxygen atoms, as well as vibrations of deformation of the angle formed between H, O and H atoms (Figure 3.17.b). Based on the wavelength at which these vibrations occur, the main chemical compounds present in the structure of the analyzed sample can be identified [5].

a)   b)

Figure 3.17. Vibrations produced in the water molecule by the absorption of the IR spectrum: a) elongation or compression vibration; b) deformation vibration.

The analysis of the geopolymer samples was performed by attenuated total reflection transmission using the Bruker Hyperion 1000 micro-FTIR spectrometer (Figure 3.18) which is equipped with a 15X objective. The determination of the chemical structure of the obtained geopolymers was performed between 4000 cm^{-1} and 600 cm^{-1} with a resolution of 4 cm^{-1} at a scanning frequency of 10 KHz through an aperture size of 6 mm and a number of 64 scans. The reflection FTIR spectrum of the samples shows several maxima contained in the vibration bands of the chemical bonds in the present groups. The spectra were analyzed using OPUS 65 Bruker software (Bruker, Germany) in order to study, in particular, the groups formed between Si, Al, H and O.

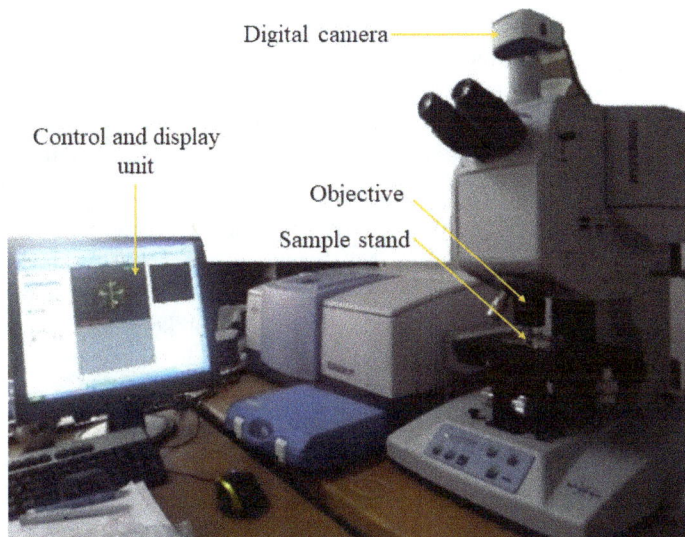

Figure 3.18. IR Spectrometer type Bruker Hyperion 1000 [20].

The ATR lens from the objective offers optimum infrared light throughput with a clear view on the studied samples. In the case of geopolymers, once the sample has been positioned in the visual image, the IR detection is enabled, while the ATR crystal made of germanium is positioned automatically. Furthermore, the acquisition system collects the data, and the FTIR spectrum is displayed [20].

3.3.12 Pore size relative distribution

Nuclear magnetic resonance (NMR) is a completely non-perturbative and non-destructive technique for investigating materials [22]. It is based on the control/manipulation of nuclear spines by means of an NMR spectrometer which is composed of a magnet, a radio frequency (RF) unit and a component that ensures the homogeneity of the magnetic field (Figure 3.19).

NMR relaxometry techniques are based on the polarization of nuclear spines in the presence of an external magnetic field and then bringing them into unbalanced conditions, by applying radio frequency pulses. The return in equilibrium conditions of the transverse and longitudinal components of the magnetization is characterized by the transverse (T_2) and longitudinal (T_1) relaxation times, respectively.

Nuclear magnetic resonance (NMR) is one of the most reliable techniques for investigating matter, being applied in the study of both liquids and solids and gases. Only plasma, the fourth state of aggregation of matter, has escaped (so far) MRI investigations. Unlike other matter investigation techniques, nuclear magnetic resonance is completely non-perturbative and non-destructive. Samples investigated by MRI can then be used in other experiments.

The best-known application of the nuclear magnetic resonance phenomenon is in medicine, namely MRI tomography (or MRI imaging) but nuclear magnetic resonance can be just as useful in chemistry, biology, materials science, soil science, oil extraction. In chemistry, MRI spectroscopy in high fields is most often known, but diffusimetry or MRI relaxometry has also proved to be particularly useful. In oil extraction, the study of soils, porous environments, and the migration of molecules through them are often applied diffusion techniques and NMR relaxometry in low fields.

Nuclear magnetic resonance techniques can be so varied that their number is limited only by the imagination and skill of the experimenter. I could say that a single NMR spectrometer allows the design of more than 1001 distinct experiments, unprecedented in other matter investigation techniques. To support this statement, Figure 1 shows the distance and time scales that can be tested by various NMR techniques. It is thus observed that by using different nuclear magnetic resonance techniques 12 orders of magnitude can be covered in distance and the same number in time.

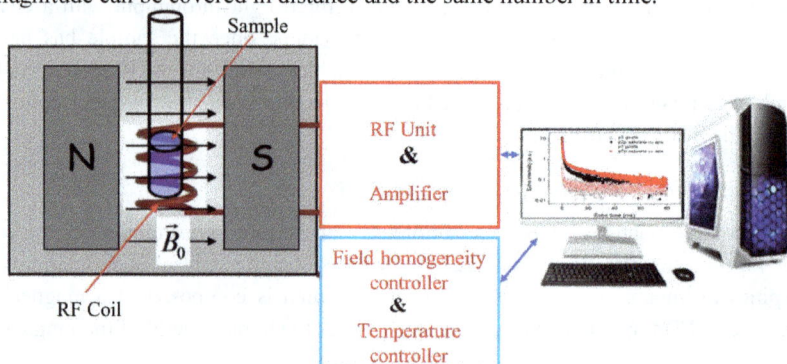

Figure 3.19. Schematic representation of an NMR spectrometer [22].

In the case of porous materials, saturated with a liquid, the transverse and longitudinal relaxation times depend on the pore size, the nature of the liquid saturating the respective pores and the nature of the pore surface. This dependency is given by the relation eq. (3.1):

$$\frac{1}{T_i} = \rho_i \frac{S}{V} = \rho_i \frac{3}{R} \ (i = 1, \ 2) \tag{3.1}$$

Equation 2.1 is valid if the contribution of the volume component of the relaxation rate is neglected, a condition always encountered in the case of porous materials containing magnetic impurities. In this equation, the constant is called relaxivity and depends on the nature of the surface of the porous material. S/V represents the surface-volume ratio of the pores, and R the pore radius, in the case of the spherical approximation.

As can be seen, eq. 2.1 provides a direct link between pore size and transverse or longitudinal relaxation time, if the relaxivity constant is known. This constant can be obtained by calibration with other techniques (Mercury or Nitrogen intrusion porosimetry). Even if the relaxivity is not known, the MRI relaxometry technique can be used to extract relative values of pore size and monitor their evolution over time or under the action of factors. In the research reported in the thesis, the relative values of pore size were monitored by providing information on the relative pore distribution of geopolymers.

In order to highlight the residual activation solution and the influence produced by the reinforcing particles on the relative distribution of pores, nuclear magnetic resonance was used which is based on manipulating the spin direction of atoms in water molecules confined in the pores of the obtained geopolymers.

In other words, when the equipment produces an electromagnetic impulse, the spin direction of the nuclei in the atoms will be oriented to a certain degree relative to the electromagnetic field, this orientation produces a perturbation of the field, which the equipment converts into a value in depending on the technique used (Figure 3.20).

In the case of porous media with magnetic impurities, it is necessary to reduce the diffusion effects on transverse relaxation measurements [23]. Carr-Purcell Meiboom-Gill (CPMG) is a technique for measuring transverse relaxation times that reduces diffusion effects. This technique can be used in combination with low field equipment (Figure 3.21) to reduce the influence of internal gradients [24]. In this research, the porosity assessment was performed on cylindrical samples with a diameter of 6 mm and a length of 10 mm at different ages, as well as after immersion in water for a week, in order to fill the open pores with liquid.

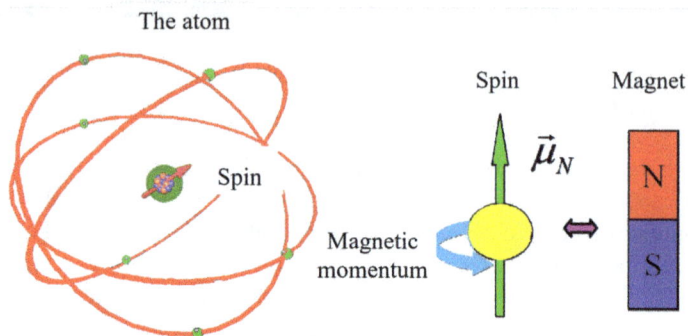

Figure 3.20. The gyroscopic model of the nuclear spin with the associated magnetic moment [22].

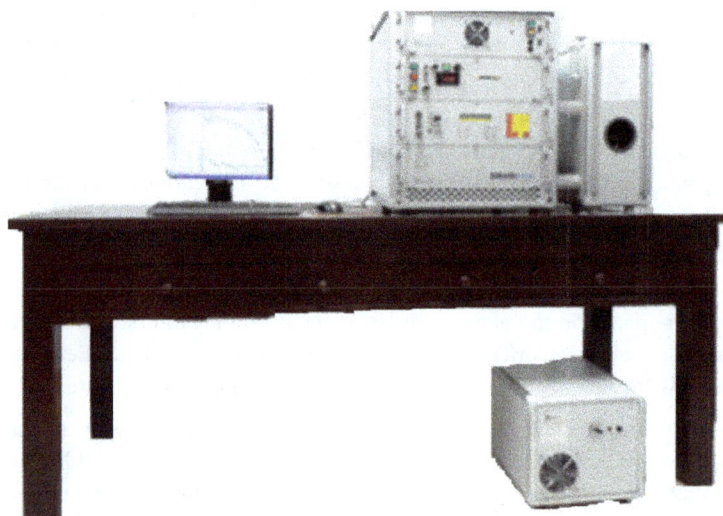

Figure 3.21. STELAR SMARtracer - Fast Field Cycling NMR relaxometer [25].

The porosity and pore structure of geopolymers can be determined by mercury intrusion porosimeters (MIP), gas absorption tests (BET), nuclear magnetic resonance (NMR) and X-ray computed tomography (X-CT). However, only Nuclear Magnetic Resonance (NMR) Relaxometry allows the detection of the porous structure in a completely non-disturbing way. In the case of geopolymers, NMR relaxometry can be used to monitor the relaxation of protons belonging to molecules in liquids trapped inside the porous structure and thus to have access to the pore size distribution. This monitoring can take place even during the polymerization process [26].

3.3.13 Methods used for physical-mechanical evaluation

In order to study the mechanical properties of geopolymer samples, their specific structure (porous material with a high concentration of defects near the area of the matrix-aggregate interface) must be considered. The breaking process, specific to fragile materials, of geopolymer samples takes place in three stages:

• initiation of cracks;

• crack propagation;

• growth and development of cracks.

In oxide materials, under the action of an external force, cracks appear that do not propagate in a straight line but follow a sinusoidal direction around the aggregates or around the constituents present in the structure. The direction of propagation is strongly influenced by the presence of aggregates or pores, as they can block or deflect them.

3.3.13.1 Compressive strength evaluation

Compressive strength is one of the main criteria for verifying the quality of a geopolymer sample. The value obtained depends on the test conditions, the dimensions and the geometric shape, the manufacturing and storage conditions, as well as the loading speed of the analyzed sample. Due to the friction between the test surfaces and the plates of the test equipment, tangential stresses occur at the boundary of the geopolymer-metal separation which prevents the transverse deformation of the specimen. Destruction of the sample occurs by detaching its lateral surfaces according to planes inclined at about 30° from the vertical (Figure 3.22). The formation of cracks can be observed, in the form of fluctuations and in the loading / displacement graphs (Figure 3.23).

In order to determine the compressive strength of the geopolymer samples, the requirements of Standard C109 / C 109M-07 - Standard test method for the compressive strength of hydraulic cement mortar have been complied with. The specimens were tested at various ages, 7 days, 28 days and 90 days, which are calculated as the time period between the time of demodeling (removal of specimens from molds or molds immediately after the completion of the drying process) and the time of testing.

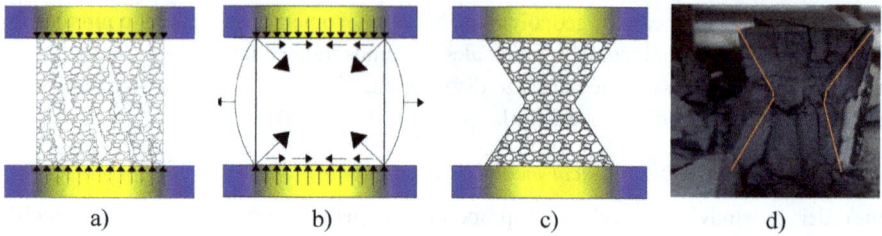

a)   b)   c)   d)

Figure 3.22. The breaking behavior, during compression, of the geopolymer samples, with friction between the sample surfaces and the plates. a) initial state; b) schematic representation of tensions; c) theoretical fracture; d) experimental fracture [27].

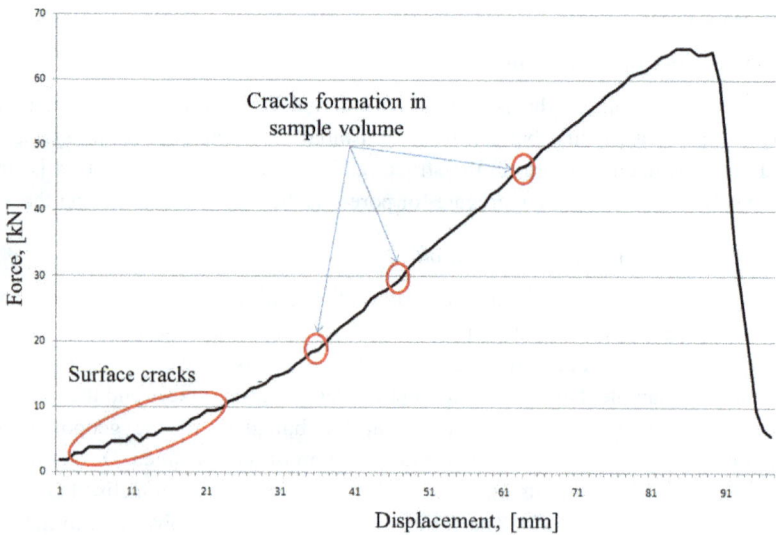

Figure 3.23. Load/displacement graph of a sample during compression [27].

In order to eliminate some factors that may influence the values obtained for the compressive strength, it is necessary to establish procedures for handling and preparing the test pieces. In this case the preparation was carried out as follows:

• clean the faces of the specimen with a clean cloth to remove any dust particles or other material. Wipe the load-bearing surface of the test machine, the load-bearing plates and the fastening device with a clean cloth and position the test piece in the test machine so that the load is applied to one of its faces, perpendicular to the direction of casting;

• the test piece was placed so that the load was applied over the entire width of the contact faces with the plates. When using the load-bearing plates and the fastening device, place a load-bearing plate on the upper surface of the fastening device with its longitudinal axis parallel to the indicator arrow ensuring the contact of the whole surface. The test piece was positioned in the fixing device, between the columns, with the longitudinal axis perpendicular to the arrow and another load-bearing plate was placed on the test piece, parallel to the lower load-bearing plate. Carefully center the fastener assembled on the lower deck of the test machine;

• the task was applied by progressive increase until the rupture occurred;

• the mechanical compressive strength of each specimen was recorded, with an accuracy of 0.05 N/mm². The average value was calculated with an accuracy of 0.1 N/mm².

Compressive strength was calculated from the relation (3.2):

$$Rc = \frac{P}{A}, [N/mm^2] \tag{3.2}$$

where: P - breaking force in N; A - the area of the section in mm².

Controls C56 3000 kN hydraulic press (Figure 3.24) used to perform compressive strength tests of geopolymer samples. C56 3000 kN (Controls Group, Italy) is a device that uses a Digimax Plus unit for data acquisition.

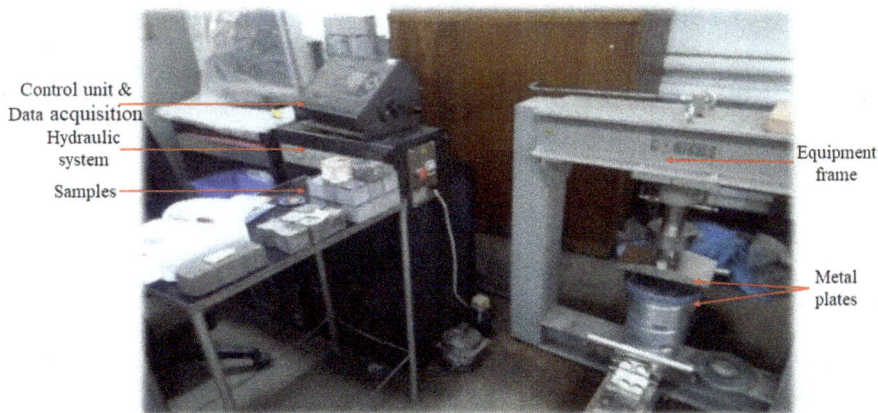

Figure 3.24. Hydraulic press type Controls C56 3000 kN [28].

The maximum value, expressed in kN, corresponds to the sudden drop in load during the test. At that time the equipment considers that the breaking of the sample has taken place and the test is completed.

3.3.13.2 Flexural strength evaluation

The bending strength was tested, according to Standard SR EN 196-1, by applying a force at the center point on the upper surface of a test specimen, when the lower surface is placed on two supports (Figure 3.25).

The value of the bending strength obtained in this way is influenced both by the shape of the cross-section of the specimen and by the ratio between length and thickness.

The Controls L15 bending strength test equipment allows the bending strength of samples to be recorded with an accuracy of 0,05 N/mm^2, by means of three points bending method (Figure 3.26).

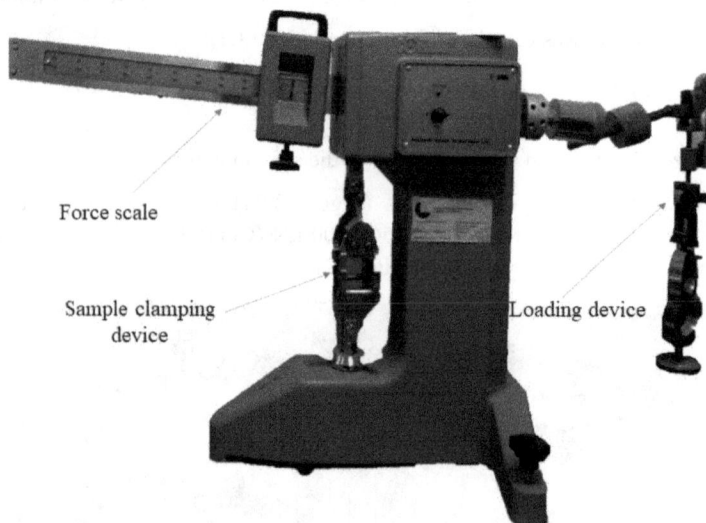

Force scale

Sample clamping device

Loading device

Figure 3.25. Bending strength test machine type Controls L15.

The specimens were tested at various ages, 7 days, 28 days, and 90 days in order to analyze the evolution over time of bending strength.

The bending strength, f, was calculated from the relation (3.3):

$$f = 1,5 \cdot \frac{F \cdot l}{b \cdot d^2}, [N/mm^2] \tag{3.3}$$

where: F is the maximum force applied to the specimen, (N); l represents the distance between the axes of the support rollers, (mm); b is the width of the test piece, (mm); d represents the thickness of the test piece, (mm).

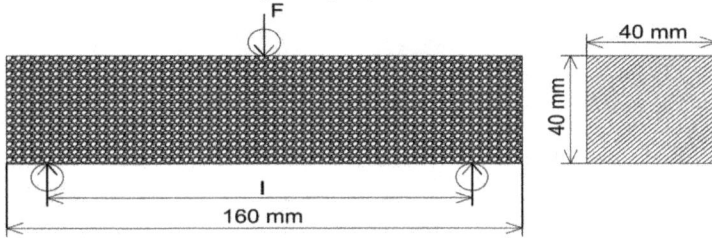

Figure 3.26. Schematic representation of three-points bending strength testing with central loading.

Controls L15 offers the possibility to evaluate the bending strength of geopolymer specimens with dimensions of 40x40x160 mm^3 by the method of the three points with central loading.

3.3.14 Evaluation of thermal behavior

Simultaneous Thermal Analysis (TG-DTA) consists in determining the evolution of the mass of geopolymer samples by thermogravimetric analysis (TGA - Thermogravimetric Analysis) simultaneously with the study of phase transformations by differential thermal analysis. The study of materials by TG-DTA highlights the thermal stability of geopolymers by quantitative determination of the types of compounds that have volatilized in a certain temperature range.

In the case of geopolymers, the TG-DTA technique can highlight the phase transformations or mass losses that occur in the analyzed temperature range. The analyzes were performed using Linseis STA PT-1600 equipment (Figure 3.26).

TG-DTA is a thermal analysis technique, by which the difference in the amount of heat required to increase the temperature of the analyzed sample is compared with that of a control sample, obtaining a function of temperature over time [29]. This technique is based on the fact that when the studied sample undergoes a certain transformation, it absorbs a certain amount of heat, in case of an endothermic reaction, or releases heat, in case of an exothermic reaction, thus a difference is detected compared to the reference sample (Figure 3.27). For example, when a sample changes from solid to liquid state, this transformation will require an additional heat input compared to the reference sample. At the same time, the decomposition of some compounds from the material or its oxidation can cause a loss or increase in mass of the sample.

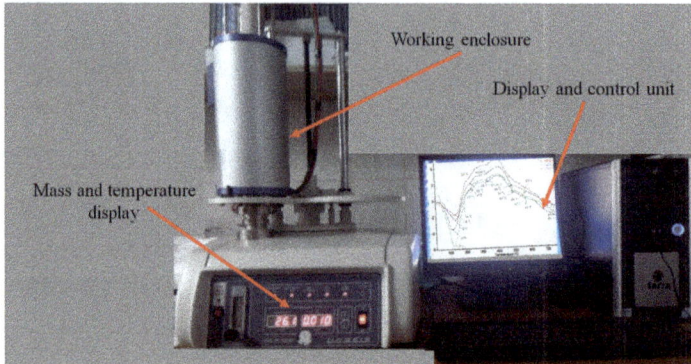

Figure 3.26. Linseis STA PT-1600 thermobalance [30].

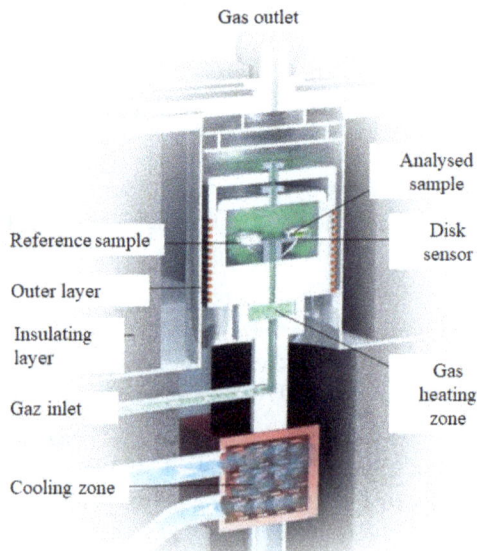

Figure 3.27. Thermobalance operating principle [31].

As the use of geopolymers in industrial applications is constrained by the characteristics of the material, as well as the complexity of the method of production, in order to identify suitable applications for these materials, it was established to evaluate the chemical, structural, physical-mechanical and thermal characteristics by specific study methods. of oxide materials.

In conclusion, the characterization of geopolymers and the establishment of the dependence between the chemical characteristics and the structural, physical-mechanical and thermal characteristics of geopolymers, took into account different types of investigations.

References

1.   Ingham, J.P. Bricks, terracotta, and other ceramics. In *Geomaterials Under the Microscope*; Elsevier, 2013; pp. 163–170.

2.   Tsega, E. Effects of Firing Time and Temperature on Physical Properties of Fired Clay Bricks. *Am. J. Civ. Eng.* 2017, *5*, 21. https://doi.org/10.11648/j.ajce.20170501.14.

3.   Etuve cu convectie naturala model ED, producator BINDER | Analitic Laboratory Available online: http://www.analiticlaboratory.ro/etuve-cu-convectie-naturala-model-ed/ (accessed on Sep 17, 2020).

4.   Sieve Shaker AS 200 control - RETSCH - precise sieve analysis Available online: https://www.retsch.com/products/sieving/sieve-shakers/as-200-control/function-features/ (accessed on Feb 12, 2020).

5.   Vibratory Sieve Shaker AS 200 basic - RETSCH - short sieving times Available online: https://www.retsch.com/products/sieving/sieve-shakers/as-200-basic/function-features/ (accessed on Sep 20, 2020).

6.   Adam Equipment PW254 Laboratory Balance Balance - Precision Weighing Balances Available online: http://balance.balances.com/scales/615/ (accessed on Sep 20, 2020).

7.   Adam Equipment, available online: https://www.johnmorrisgroup.com/Content/ Attachments/134958/11700-58.pdf (accessed on Sep 20, 2020).

8.   Asphalt Prism Shearbox Compactor PReSBOX®, Asphalt/bituminous mixture testing equipment, Controls Available online: https://www.controls-group.com/eng/asphaltbituminous-mixture-testing-equipment/asphalt-prism-shearbox-compactor-presbox_.php (accessed on Oct 8, 2020).

9.   SEM-EDS - MooreAnalytical, available online: https://www.mooreanalytical.com/

sem-eds/ (accessed on Feb 12, 2020).

10.   The structure of anorganic compounds Available online: https://alili2001.files.
wordpress.com/2014/12/m04_chimorganica.pdf (accessed on Feb 20, 2020).

11.   How Does XRF Work? - Handheld XRF Analyzer Spectrometer, X-ray Fluorescence
Analyzer, PMI Gun, XRF Scanner Instrument, XRF Tester Device, Portable XRF Analyzer
Machine | Bruker Available online: https://www.bruker.com/products/x-ray-diffraction-and-
elemental-analysis/handheld-xrf/how-xrf-works.html (accessed on Sep 20, 2020).

12.   S8 TIGER - X-ray Fluorescence, XRF, elemental analysis | AXS Bruker | Bruker
Available online: https://www.bruker.com/products/x-ray-diffraction-and-elemental-
analysis/x-ray-fluorescence/s8-tiger.html (accessed on Oct 8, 2020).

13.   Axioscope for Materials - Microscope for Research and Routine Available online:
https://www.zeiss.com/microscopy/int/products/light-microscopes/axioscope-5-for-
materials.html (accessed on Oct 8, 2020).

14.   Forcipol 2V Grinder & Polisher Available online: https://www.kemet.
co.uk/images/downloads/Forcipol2V.pdf (accessed on Oct 8, 2020).

15.   Lin, C.L.; Chen, F.S.; Twu, L.J.; Wang, M.J.J. Improving SEM inspection
performance in semiconductor manufacturing industry. *Hum. Factors Ergon. Manuf.*
2014, *24*, 124–129. https://doi.org/10.1002/hfm.20360

16.   The Applications and Practical Uses of Scanning Electron Microscopes - ATA
Scientific Available online: https://www.atascientific.com.au/sem-imaging-applications-
practical-uses-scanning-electron-microscopes/ (accessed on Sep 20, 2020).

17.   Quanta Scanning Electron Microscope | Thermo Fisher Scientific Available
online: https://www.fei.com/products/sem/quanta-sem/#gsc.tab=0 (accessed on Oct 8,
2020).

18.   Scanning electron microscopy, Available online: http://www.mdeo.
eu/MDEO/Studenti/Docs/SEM_seminar_2012.pdf (accessed on Sep 20, 2020).

19. EERIS: Infra Public Profile Available online:
https://eeris.eu/index.php?&sm=module.org.erris.app.infra&ddpN=3245192760&we=a5
ba74f6d75889ea8c62a266f3e019f6&wf=dGFCall&wtok=2039ba1aefe53d1c7fa9577468
362ebe572f863b&wtkps=JY1BEoMgEAT/snctAVFZXwMIiBoUiGWsVP6eaG596OmR
WOM7I0PIfoA+IycI2zQucWfxZeN8Ru0HoRrmJh25WwM5m1S5rrD7OfJQkKSoC/Sat
gg+2CT/oRqBUNreTYTjOS/5YkYRFOFKKEMY7bTlUkktuKiktkrXmlllеT/tMDc0CI9
12BdTrsmVJiWfS7lt5f0F/ecL&wchk=f5aa64048abe4aa86a62fae67a81102d279106c4
(accessed on Oct 9, 2020).

20.   Hyperion brochure, Available online: https://www.bruker.com/fileadmin /user _upload/8-PDF-Docs/OpticalSpectrospcopy/FT-IR/Hyperion/Brochures/HYPERION _Brochure_EN.pdf (accessed on Sep 20, 2020).

21.   X'Pert³ MRD | XRPD | XRD Materials Research System | Malvern Panalytical Available online: https://www.malvernpanalytical.com/en/products/product-range/xpert3-range/xpert3-mrd (accessed on Oct 8, 2020).

22.   Ardelean, I. *Rezonanţa magnetică nucleară pentru ingineri*; ISBN 978-973-662-905-1.

23.   Ardelean, I.; Farrher, G.; Mattea, C.; Kimmich, R. NMR study of the vapor phase contribution to diffusion in partially filled silica glasses with nanometer and micrometer pores. In Proceedings of the Magnetic Resonance Imaging; Elsevier Inc., 2005; Vol. 23, pp. 285–289.

24.   Pop, A.; Bede, A.; Dudescu, M.C.; Popa, F.; Ardelean, I. Monitoring the Influence of Aminosilane on Cement Hydration Via Low-field NMR Relaxometry. *Appl. Magn. Reson.* 2016, *47*, 191–199. https://doi.org/10.1007/s00723-015-0743-7

25.   Equipment – NMR diffusometry and relaxometry laboratory Available online: https://nmr.utcluj.ro/equipment/ (accessed on Sep 20, 2020).

26.   Pop, A.; Ardelean, I. Monitoring the size evolution of capillary pores in cement paste during the early hydration via diffusion in internal gradients. *Cem. Concr. Res.* 2015, *77*, 76–81. https://doi.org/10.1016/j.cemconres.2015.07.004

27.   Burduhos Nergis, D.D.; Vizureanu, P.; Corbu, O. Synthesis and characteristics of local fly ash based geopolymers mixed with natural aggregates. *Rev. Chim.* 2019, *70*.

28.   Testing systems for determining the mechanical properties of concrete and cement Available online: http://www.abmbv.nl/files/controls_compression_and_flexural_testing _machines_2013.pdf (accessed on Feb 16, 2020).

29.   Nergis, D.D.B.; Abdullah, M.M.A.B.; Sandu, A.V.; Vizureanu, P. XRD and TG-DTA study of new alkali activated materials based on fly ash with sand and glass powder. *Materials (Basel).* 2020, *13*. https://doi.org/10.3390/ma13020343

30.   STA PT 1600, TGA PT1600, DTA PT1600, Available online: http://www.gammadata.se/assets/Uploads/STA-TGA-DTA-1600-1.pdf (accessed on Oct 8, 2020).

31.   KV, K.; SR, A.; PR, Y.; RY, P.; VU, B. Differential Scanning Calorimetry: A Review. *Res. Rev. J. of Pharmaceutical Anal.* 2014, *3*, 11–22.

4. Design and Development of Geopolymers Based on Thermal Power Plant Ash

4.1. Raw materials characterization

Any material rich in silicon and aluminum that can be dissolved by an alkaline solution is a source of raw material for geopolymers manufacturing. Globally, multiple mineral wastes with potential for geopolymerization have been identified, such as: thermal power plant ash [1,2], red sludge [3–6], blast furnace slag [7,8] etc. After the chemical reaction between the solid raw material (waste) and the alkaline activator (a solution of silicates and hydroxides) a total inorganic material with a structure similar to that of zeolites is obtained [9–11]. In this research, the thermal power plant ash was used in order to obtain the matrix and two types of reinforcement particles (glass powder and sand particles).

4.1.1. Thermal power plant ash characterization

At national level (in Romania), there are large areas occupied by industrial waste resulted from burning coal in urban thermal power plants. The thermal power plant ash used to carry out the geopolymer tests comes from the dupmsides of the company CET II Holboca Iași, which occupied an area of approximately 40 hectares in 2017 [12,13]. The performance of this type of raw material in geopolymers manufacturing is strongly influenced by the concentration of elements that lead to the formation of the Si-O-Al structure.

The thermal power plant ash is a by-product of coal combustion in thermal power plants or in cityes heating power plants. Depending on the collecting method, it is divided into two categories: fly ash and bottom ash [14–16]. The SEM micrograph of the power plant ash (Figure 4.1) presents spongy particles (particles with porous structure) of irregular shape and spherical particles.

The quantitative chemical composition (Table 4.1) of the power plant ash used in this research was analyzed after the drying and particle size separation steps.

Table 4.1. Oxide chemical composition of power plant ash.

Oxide	SiO_2	Al_2O_3	Fe_xO_y	CaO	K_2O	MgO	TiO_2	Na_2O	Oth.
[%], wt.	47.8-48.0	28.8-29.0	10.0-10.2	6.2-6.4	2.0-2.5	2.0-2.1	1.3-1.4	0.6-0.8	<0.1

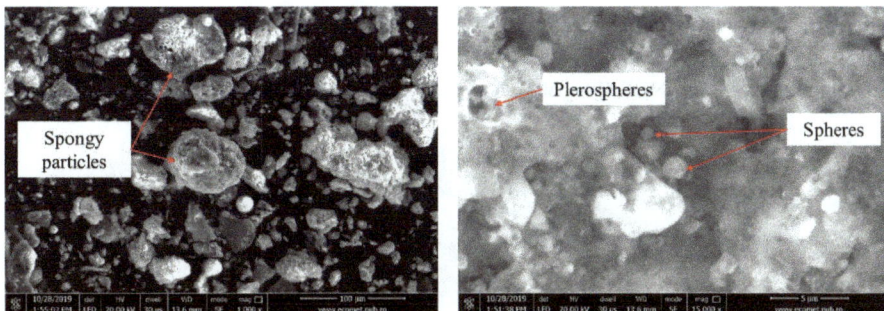

Figure 4.1. Morphology of power plant ash particles.

According to the ASTM C618-92a standard, power plant ash belongs to class F, because the sum of silicon, aluminum and iron oxides is higher than 70% (ec. (4.1)).

$$SiO_2 + Al_2O_3 + Fe_2O_3 = 47.8\ \% + 28.6\ \% + 10.2\ \% = 86.6\ \% \qquad\qquad (4.1)$$

4.1.2. Glass powder characterization

Another mineral waste that results in large quantities, both from the food industry and from the civil construction industry is glass. This inert material does not decompose naturally, producing proven negative effects on the environment following storage in landfills [17–19]. Therefore, the use of glass waste in the manufacture of environmentally friendly materials has become a global concern [20–22]. Due to the capacity of geopolymer paste to incorporate different types of particles, the introduction of glass powder in the composition of these materials can be done by simple methods (mixing in dry and wet state).

By comparing the chemical composition of glass powder (Table 4.2) with that of thermal power plant ash (Table 4.1), the glass powder has a much higher content of SiO_2, CaO and Na_2O, but much lower in Al_2O_3. However, according to several studies, the glass powder possess the capacity to reacts in alkaline environments, thus contributing to the geopolymerization reaction.

Table 4.2. Oxide chemical composition of glass powder.

Oxide	SiO_2	Al_2O_3	Fe_xO_y	Na_2O	CaO	MgO	Oth.
[%], wt.	70-71	1.5-2	0.8-1	12-14	9-11	2-3	< 0,1

The glass powder (Figure 4.2) used as a reinforcing element in geopolymer samples contains only particles smaller than 10 μm (SR EN 933-1 / 2012) and is obtained by grinding by S.C. New NCR Recycling S.R.L.

Figure 4.2. 10X optical micrograph of glass powder.

The use of glass powder in the development of geopolymers is encouraged by its positive influence on the mechanical properties of the material and the social advantage offered by recycling mineral waste with a negative impact on the environment [23–25].

4.1.3.   Characterization of the natural aggregate

In order to improve the mechanical properties of geopolymers based on thermal power plant ash, different quantities or types of aggregates can be added to the composition. In addition to the use of waste, another category of reinforcing elements studied worldwide are natural aggregates (in this case sand). Depending on the chemical composition (Table 4.3) and the geometric particularities of the particles, by introducing them into the geopolymer matrix, an increase of compressive strength up to 150% can be obtained [26].

The quantity and type of aggregate used are chosen depending on the particle size distribution of the sand, because it can affect the homogeneity of the samples, but also their porosity. For this purpose, a suite of sieves with the dimensions shown in Table 4.4 was used.

Table 4.3. Oxide chemical composition of sand particles.

Oxide	SiO_2	Al_2O_3	Fe_xO_y	Oth.
[%], wt.	98.0 – 99.0	0.3-0.5	0.2-0.3	< 0.1

Table 4.4. Particle size distribution of the natural aggregate [26].

Sieve	Dimension, [mm]	Sample weight, [g]	Retain, [%]	Pass, [%]
1	4.00	0	0.00%	100.00%
2	3.75	10.90	10.90%	89.10%
3	2.00	14.00	14.00%	75.10%
4	0.21	18.70	18.70%	56.40%
5	0.12	36.70	36.70%	19.70%
Jar	0	19.70	19.70%	0.00%

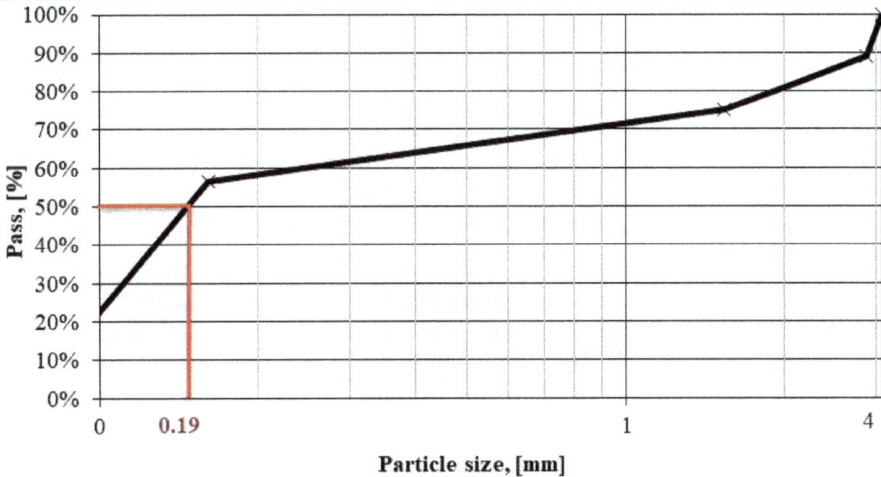

Figure 4.3. Particle size distribution of aggregates used [26].

The granulometric characteristics evaluation by sieving (SR EN 933-1 / 2012) was performed after drying the aggregates (the samples were heated and mainteined at 120 °C until a constant mass of the raw material is obtained) in order to reduce the measurement errors caused by the sticking or adhesion of fine sand particles to the sieves surface. According to the particle size distribution shown in Figure 4.3, the raw material used contains approximately 50% of the particles with a diameter less than or equal to 0.19 mm. Therefore, the type of sand used belongs into the class of aggregates 0/4 because all particles pass through the sieve with a mesh size of 4 mm (SR ISO 3310-3).

4.2. The geopolymers designing

A geopolymer is mainly composed of two components, the base material (the solid component) and the alkaline activator (the liquid component). The majority component is the base material, it must be rich in silicon and aluminum and can be a natural mineral (clays, kaolin) or industrial waste (power plant ash, red mud, slag etc.). There are several factors that influence the decision of choosing a raw material to obtain a geopolymer, these factors being closely related to the cost or availability of the source, as well as the resulting geopolymer application.

Therefore, the main design criterion of this material was the compressive strength. The design of the four types of geopolymers, within the experimental research, was made based on the influence they produce, both the proportion and the type of reinforcing elements, on the compressive strength of geopolymers. Accordingly, a series of preliminary experiments were performed for the elaboration/obtaining of some types of geopolymers, with different proportions of glass powder or sand. The samples obtained were in classes B150 (equivalent to C8/10) and B250 (equivalent to C16/20) of Portland cement-based concrete, according to the results obtained by testing the samples according to the requirements of standard C109/C109-07 - Standard test methods for compressive strength of hydraulic cement mortar.

Geopolymers containing several constituents in the solid component must be mixed in the dry state before contact with the activation solution. The main purpose of dry mixing is to increase the homogeneity of the final structure. At the same time, activating the geopolymer with a multicomponent solution involves mixing these components before introducing them into the solid component.

Because the characteristics and properties of geopolymers depend on multiple factors (Table 4.5), it is essential to establish the optimal parameters specific to the raw material, the activation solution, the drying step and the mixing step. The experimental methodology presented above was established in order to obtain geopolymeric materials that meet requirements comparable to those of Portland cement-based construction materials. In accordance to preliminary experiments and taking into account the literature [27–33] , the following parameters were established:

- Coal ash particle diameter: ≤80 μm;
- Relative humidity of raw material: 0%;
- Silicon and aluminium oxides concentration: ≥75%; Calcium content: <10% (Figura 3.4);
- Curing temperature: 70 °C;
- Curing time: 8, 16 sau 24 de ore;
- Solid to liquid ratio (by mass): 1 (ec. (4.2));

$$\frac{\text{Solid component (Coal ash} \pm \text{reinforcing elements), [g]}}{\text{Liquid component (Sodium silicate + Sodium hydroxide, 10 M), [g]}} = 1 \qquad (4.2)$$

- Sodium silicate to sodium hydroxide ratio (by mass): 1.5 (ec. (4.3));

$$\frac{\text{sodium silicate solution, [g]}}{\text{sodium hydroxide solution, 10M, [g]}} = 1{,}5 \qquad (4.3)$$

- Molar concentration of sodium hydroxide solution: 10 M.

Table 4.5. Factors influencing the obtaining of geopolymers.

Raw materials	Particle size distribution
	Humidity
	Chemical compozition
Activator solution	Solid to liquid ratio
	Sodium silicate to Sodium hydroxide ratio
	Sodium hydroxide solution ratio
Drying/curing stage	Curing time
	Curing temperature
Mixing stage	Mixing time
	Mixing speed

Also, the effect of replacing the coal ash in certain percentages with reinforcement particles (Table 4.6) with the following characteristics will be studied:

- glass with particles diameter lower than 10 μm;
- sand with particles diameter lower than 4 mm.

Table 4.6. The components o the obtained geopolymers.

Sample	Liquid component		Solid component		
	Na₂SiO₃, [%]	NaOH, [%]	Coal-ash, [%]	Glass powder, [%]	Sand, [%]
100FA	60	40	100	0	0
70FA_30PG	60	40	70	30	0
30FA_70S	60	40	30	0	70
15FA_15PG_70S	60	40	15	15	70

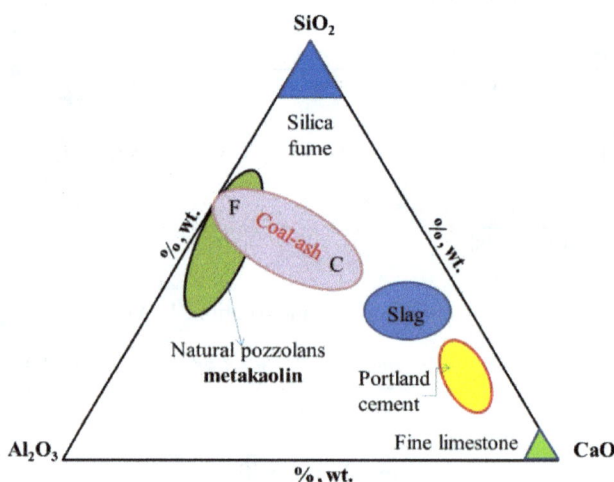

Figure 4.4. Ternari digram of the main compound from geopolymers [34–36].

During the geopolymerization process, minerals rich in aluminum and silicon pass through several phases (see Figure 1.1). In the first phase they are dissolved by the alkaline solution forming a gel whose viscosity is given by the ratio between solid and liquid. In the second phase the reorganization of the molecules takes place, after the elimination of water, and the beginning of the material hardening following the formation of a tetragonal structure of Si-O-Al containing polysialate, polysialate-siloxo or polysialate-disiloxo bonds (see Table 1.1).

4.3. Obtaining of coal-ash based geopolymers

The moisture of the power plant ash can influence the alkalinity of the activator, consequently the dissolution capacity of the solution can be diminish. In order to eliminate this disturbing factor (humidity), a drying stage of the coal ash was used after its sampling.

The drying of the coal-ash is carried out in rooms with controlled heating without the use of ventilation to reduce the loss of fine particles from the material. The process was performed at a temperature of 120 °C using a Binder Stove which has the possibility to adjust with high precision the temperature (1 °C) and the time (1 minute). The determination of the optimal drying period was performed experimentally by checking the mass loss over time of a predetermined amount of ash. The ash was considered to be dry when the mass value was constant at 3 consecutive measurements performed every 10 minutes [37].

4.3.1. Solid component preparing

Overall, geopolymers made of fine particles will acquire, after hardening, superior mechanical properties and a denser microstructure compared to those obtained from larger particles [38,39]. Therefore, the particle size distribution has a significant effect on the microstructure and physical properties.

The granulometric separation step was performed immediately after the drying of the ash and had as main purpose the separation of the powder by granulometric classes and the elimination of certain impurities.

The powder was separated using the Sieve Shaker AS 200 vibration shaker [40] through several sieves, as follows: sieve 1 - 2.00 mm, sieve 2 - 0.21 mm, sieve 3 - 0.12 mm, sieve 4 - 0.08 mm followed by the collecting container.

4.3.2. Liquid component preparation

The activation of the raw material is one of the most important factors in the production of a geopolymer, because the activator influences the precipitation and crystallization of the aluminum and silicon species existing in the solution. The OH- groups, from it, act as a catalyst for reactivity, and Na+ metal cations serve to balance the tetrahedral Si-O-Al structure.

When power plant ash or another material rich in aluminum and silicon is mixed with the alkaline solution, the glass component is quickly dissolved, but there is not enough time and space for the formed gel to pass into a completely crystalline structure. Consequently, the resulting material has a mixed amorphous and crystalline structure [41].

In order to obtain geopolymers based on power plant ash, a mixture of sodium hydroxide and sodium silicate was used as activator, in this research.

4.3.2.1. Sodium hydroxide solution (NaOH)

The concentration and molarity of the NaOH solution strongly influence the final properties of the geopolymers. High concentrations of NaOH solution provide high resistance to early stages of the reaction. NaOH-activated geopolymers have high crystallinity, therefore, they have better stability in acidic or sulphate environments [42].

The NaOH solution was prepared at a molar concentration of 10 by dissolving high purity NaOH flakes (98%) in distilled water at least 24 hours before use.

4.3.2.2. Sodium silicate solution (Na$_2$SiO$_3$)

Sodium silicate is obtained by fusing sand with sodium hydroxide or sodium carbonate at temperatures above 1300 °C. In geopolymers, it is rarely used as an independent activator because it does not have a sufficiently high dissolving capacity to initiate the geopolymerization reaction.

In this case, the sodium hydroxide solution was mixed with a commercially purchased high-purity sodium metasilicate (Na$_2$SiO$_3$) solution (Scharlab SL, Spain), with a density of 1.37 g/cm^3 and a pH lower than 11.5 [43].

4.3.3.　Components mixing stage

The main factor influencing the homogeneity of the obtained samples is the mixing of the components [44,45]. The required mixing time is obtained by visual inspection of the uniformity of the paste.

The mixing of the components used in the manufacture of geopolymers was performed using the IMEC 275 powder mixer, according to the following procedure:

- the dry coal-ash was introduced into the vessel;
- glass powder/sand has been added to the vessel, according to the specification presented in Table 4.5;
- the solid components were mixed for 5 minutes, in the case of samples with reinforcing particles;
- the activation solution was added;
- both components (solid and liquid) were mixed for 10 minutes.

The mixing stage has been carried out according to the specfications presented in BS EN 196-1: 1995.

It is important that when the composition includes several solid components they should be mixed in the solid state, for a short period of time, before adding the liquid component. The introduction of the activation solution in the mixer vessel and then of the solid component leads to the decrease of the homogeneity of the final mixture and to the formation of unreacted ash areas [26,27].

4.4. The process flow of coal ash based geopolymers obtaining

The coal-ash based geopolymers were obtained according to the process flow diagram presented in Figure 4.6.

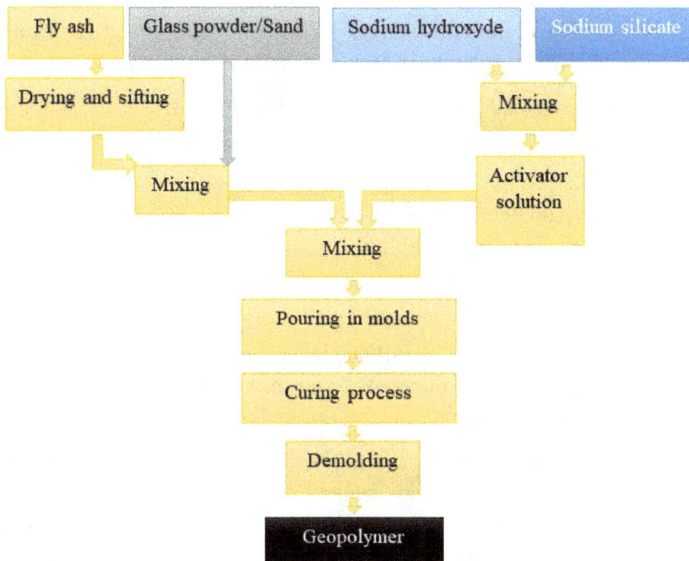

Figure 4.6. Flowchart of geopolymers obtaining.

Stage 1 consists in the preparation of the solid components by *collecting* the power plant ash from dump area, *drying* and particle size separation (*sifting*) and reinforcing particles drying.

Step 2 consists of mixing the solid components in the dry state in order to increase the homogeneity of the final mixture.

The third step is to prepare the liquid component by mixing the sodium hydroxide solution with the sodium silicate solution. The NaOH solution is prepared at the desired concentration by dissolving the NaOH flakes in distilled water at least 24 hours before mixing the solid and liquid components, in order to eliminate the influence of the exothermic reaction of sodium hydroxide solution on the geopolymerization process.

The 4th stage consists in mixing the solid components with the liquid ones with the help of an adjustable speed mixer, for 10 minutes until a homogeneous paste is obtained.

The 5th stage consists in the vibration of the samples, on a vibrating table, in order to eliminate the air bubbles introduced in the mixture during the mixing stage of the components.

The 6th stage consists in drying the samples for 8, 16 or 24 hours at 70 °C, followed by demodeling (extraction of the samples from molds) after cooling them.

The samples obtained were kept in air, at ambient temperature (22 ± 2 °C), until the test age (7 days, 28 days, 90 days etc.).

The experimental research program aimed at the design, obtaining / elaboration and characterization of geopolymers based on indigenous thermal power plant ash, considering a complex and interdisciplinary study in the field of physics, chemistry, materials science and civil engineering on oxide materials. based on silicon and aluminum.

As a result of the study of the factors influencing the obtaining of these materials and of the potential sources of local raw material, four types of geopolymers based on thermal power plant ash and reinforcement particles were designed.

To obtain the four types of materials, the following was used: (i) power plant ash, with particles smaller than 80 µm, (ii) glass powder, with particles smaller than 10 µm and (iii) sand with particles smaller than 4 m mm. The samples were dried at a temperature of 70 °C for three different periods of time: 8 hours, 16 hours and 24 hours, respectively. Alkaline activation was performed using a solution of sodium silicate and sodium hydroxide with a molar concentration of 10 (10M) at a mass ratio of 1.5. The ratio between the solid component and the activation solution, determined experimentally, is 1. The solid component of the first type of geopolymer (sample 100FA) contains 100% ash from the power plant, the second type (70FA_30PG) is made from 70% ash from thermal power plant and 30% glass powder, the third (30FA_70S) contains 30% thermal power plant ash and 70% sand, and the fourth type (15FA_15PG_70S) is made of 15% thermal power plant ash, 15% powder glass and 70% sand.

The development of these new geopolymers based on thermal power plant ash is a method of recovery of mineral waste, in order to obtain new materials with mechanical and chemical properties, comparable to those of conventional materials (based on Portland cement).

In order to determine the main characteristics of the obtained geopolymers, it was necessary to evaluate the samples from the chemical (chemical composition), structural (macrostructural, microstructural and mineralogical), physical-mechanical (setting time, relative pore distribution, compressive strength and tensile strength). bending) and thermal (thermogravimetric analysis and differential thermal analysis).

In order to improve the mechanical properties of the obtained geopolymers, the effect of introducing, as a reinforcing element, the glass particles, which come from the recycling of some wastes from the food industry, was studied.

The novelty of the thesis is based on (i) the originality of the chemical composition of the raw material and the reinforcing elements and (ii) on the methodology for obtaining geopolymers. The developed materials are obtained with minimum energy consumption and optimized in terms of mechanical properties through the activation solution and the percentage of reinforcing particles.

In order to fulfill the main objective of the experimental research, which refers to the obtaining of geopolymers by capitalizing some mineral wastes, several stages of experimental research were completed:

- identification, sampling and characterization of raw materials;
- design of geopolymers as environmentally friendly materials with appropriate properties for industrial applications;
- obtaining four types of geopolymers based on thermal power plant ash with or without reinforcement particles;
- characterization of geopolymers obtained from the chemical, structural, mechanical and thermal behavior;
- evaluation and identification of geopolymers with the best chemical, structural, mechanical and thermal characteristics.

In the first stage, several preliminary experiments were carried out which led to the identification of raw materials: power plant ash, glass powder and sand. Subsequently, the sources of mineral waste, the ash from the thermal power plant (the storage dumps of the company CET II- Holboca, Iași), as well as the glass powder (collected by RECYCLE International and ground by New NCR Reciclare SRL) were identified. , respectively of the sand (class $(0 \div 4)$ mm, provided by Ungureanu Trans SRL). Then, the raw materials taken were characterized, the results highlighting the fact that the raw materials can be used to obtain geopolymers (has the potential for geopolymerization).

The design of materials, in the form of the four types of geopolymers, had as main criterion the compressive strength of geopolymers. Therefore, the proportion and type of reinforcing elements (glass powder or sand) was established according to their influence on the mechanical properties of geopolymers.

To obtain the geopolymer samples, a production technology was designed which includes the following steps: preparation of the liquid and solid component, mixing of the components, obtaining the geopolymer paste, pouring the paste into shapes and drying / obtaining the actual four types of geopolymers.

References

1.   Coal utilization - Coal combustion | Britannica Available online: https://www.britannica.com/topic/coal-utilization-122944/Coal-combustion (accessed on Feb 8, 2020).

2.   Zain, H.; Abdullah, M.M.A.B.; Hussin, K.; Ariffin, N.; Bayuaji, R. Review on Various Types of Geopolymer Materials with the Environmental Impact Assessment. *MATEC Web Conf.* 2017, *97*, 01021. https://doi.org/10.1051/matecconf/20179701021

3.   He, J.; Jie, Y.; Zhang, J.; Yu, Y.; Zhang, G. Synthesis and characterization of red mud and rice husk ash-based geopolymer composites. *Cem. Concr. Compos.* 2013, *37*, 108–118. https://doi.org/10.1016/j.cemconcomp.2012.11.010

4.   He, J.; Zhang, J.; Yu, Y.; Zhang, G. The strength and microstructure of two geopolymers derived from metakaolin and red mud-fly ash admixture: A comparative study. *Constr. Build. Mater.* 2012, *30*, 80–91. https://doi.org/10.1016/j.conbuildmat.2011.12.011

5.   Liu, Y.; Lin, C.; Wu, Y. Characterization of red mud derived from a combined Bayer Process and bauxite calcination method. *J. Hazard. Mater.* 2007, *146*, 255–261. https://doi.org/10.1016/j.jhazmat.2006.12.015

6.   Harekrushna, S.; Subash, C.M.; Santosh, K.S.; Ananta, P.; Himanshu, S.M. Progress of Red Mud Utilization: An Overview. *Am. Chem. Sci. J.* 2014, *4*, 255–279.

7.   Prusty, J.K.; Pradhan, B. Multi-response optimization using Taguchi-Grey relational analysis for composition of fly ash-ground granulated blast furnace slag based geopolymer concrete. *Constr. Build. Mater.* 2020, *241*. https://doi.org/10.1016/j.conbuildmat.2020.118049

8.   Karakoç, M.B.; Türkmen, I.; Maraş, M.M.; Kantarci, F.; Demirbola, R.; Uiur Toprak, M. Mechanical properties and setting time of ferrochrome slag based geopolymer paste and mortar. *Constr. Build. Mater.* 2014, *72*, 283–292. https://doi.org/10.1016/j.conbuildmat.2014.09.021

9.   Yan, H.; Xue-Min, C.; Jin, M.; Liu, L.P.; Liu, X.D.; Chen, J.Y. The hydrothermal transformation of solid geopolymers into zeolites. *Microporous Mesoporous Mater.* 2012, *161*, 187–192. https://doi.org/10.1016/j.micromeso.2012.05.039

10.   Ma, L.; Han, L.; Chen, S.; Hu, J.; Chang, L.; Bao, W.; Wang, J. Rapid synthesis of magnetic zeolite materials from fly ash and iron-containing wastes using supercritical water for elemental mercury removal from flue gas. *Fuel Process. Technol.* 2019, 39–48. https://doi.org/10.1016/j.fuproc.2019.02.021

11.   Oliveira, J.A.; Cunha, F.A.; Ruotolo, L.A.M. Synthesis of zeolite from sugarcane bagasse fly ash and its application as a low-cost adsorbent to remove heavy metals. *J. Clean. Prod.* 2019, *229*, 956–963. https://doi.org/10.1016/j.jclepro.2019.05.069

12.   Annual report 2017 CET 2, available online: http://www.anpm.ro/d, accessed on Feb 12, 2020.

13.   Location report for the combustion plant with a rated thermal input of more than 50 MW - CET Iaşi 2 Holboca, available online: http://www.anpm.ro/, accessed on Dec 12, 2019.

14.   Temuujin, J.; Van Riessen, A.; MacKenzie, K.J.D. Preparation and characterisation of fly ash based geopolymer mortars. *Constr. Build. Mater.* 2010, *24*, 1906–1910. https://doi.org/10.1016/j.conbuildmat.2010.04.012

15.   Santa, R.A.A.B.; Soares, C.; Riella, H.G. Geopolymers obtained from bottom ash as source of aluminosilicate cured at room temperature. *Constr. Build. Mater.* 2017, *157*, 459–466. https://doi.org/10.1016/j.conbuildmat.2017.09.111

16.   Onutai, S.; Jiemsirilers, S.; Kobayashi, T. Geopolymer Sourced with Fly Ash and Industrial Aluminum Waste for Sustainable Materials. In *Applied Environmental Materials Science for Sustainability*; IGI Global, 2017; pp. 165–185.

17.   Blengini, G.A.; Busto, M.; Fantoni, M.; Fino, D. Eco-efficient waste glass recycling: Integrated waste management and green product development through LCA. *Waste Manag.* 2012, *32*, 1000–1008. https://doi.org/10.1016/j.wasman.2011.10.018

18.   Albino, V.; Balice, A.; Dangelico, R.M. Environmental strategies and green product development: An overview on sustainability-driven companies. *Bus. Strateg. Environ.* 2009, *18*, 83–96. https://doi.org/10.1002/bse.638

19.   Everett, J.W. Solid Waste solid waste Disposal solid waste disposal and Recycling solid waste recycling , Environmental Impacts. In *Encyclopedia of Sustainability Science and Technology*; Springer New York, 2012; pp. 9979–9994.

20.   Rivera, J.F.; Cuarán-Cuarán, Z.I.; Vanegas-Bonilla, N.; Mejía de Gutiérrez, R. Novel use of waste glass powder: Production of geopolymeric tiles. *Adv. Powder Technol.* 2018, *29*, 3448–3454. https://doi.org/10.1016/j.apt.2018.09.023

21.   Bernardo, E.; Cedro, R.; Florean, M.; Hreglich, S. Reutilization and stabilization of wastes by the production of glass foams. *Ceram. Int.* 2007, *33*, 963–968. https://doi.org/10.1016/j.ceramint.2006.02.010

22.   Dondi, M.; Guarini, G.; Raimondo, M.; Zanelli, C. Recycling PC and TV waste glass in clay bricks and roof tiles. *Waste Manag.* 2009, *29*, 1945–1951. https://doi.org/10.1016/j.wasman.2008.12.003

23.   Lu, J.-X.; Poon, C.S. Recycling of waste glass in construction materials. In *New Trends in Eco-efficient and Recycled Concrete*; Elsevier, 2019; pp. 153–167.

24.   Ali, M.M.Y.; Arulrajah, A.; Disfani, M.M.; Piratheepan, J. Suitability of using recycled glass - Crushed rock blends for pavement subbase applications. In Proceedings of the Geotechnical Special Publication; 2011; pp. 1325–1334.

25.   Mohajerani, A.; Vajna, J.; Cheung, T.H.H.; Kurmus, H.; Arulrajah, A.; Horpibulsuk, S. Practical recycling applications of crushed waste glass in construction materials: A review. *Constr. Build. Mater.* 2017, *156*, 443–467.

26.   Burduhos Nergis, D.D.; Vizureanu, P.; Corbu, O. Synthesis and characteristics of local fly ash based geopolymers mixed with natural aggregates. *Rev. Chim.* 2019, *70*.

27.   Samantasinghar, S.; Singh, S.P. Effect of synthesis parameters on compressive strength of fly ash-slag blended geopolymer. *Constr. Build. Mater.* 2018, *170*. https://doi.org/10.1016/j.conbuildmat.2018.03.026

28.   Meftah, M.; Oueslati, W.; Chorfi, N.; Ben Haj Amara, A. Intrinsic parameters involved in the synthesis of metakaolin based geopolymer: Microstructure analysis. *J. Alloys Compd.* 2016, *688*, 946–956. https://doi.org/10.1016/j.jallcom.2016.07.297

29.   Panias, D.; Giannopoulou, I.P.; Perraki, T. Effect of synthesis parameters on the mechanical properties of fly ash-based geopolymers. *Colloids Surfaces A Physicochem. Eng. Asp.* 2007, *301*, 246–254. https://doi.org/10.1016/j.colsurfa.2006.12.064

30.   Singh, B.; Ishwarya, G.; Gupta, M.; Bhattacharyya, S.K. Geopolymer concrete: A review of some recent developments. *Constr. Build. Mater.* 2015, *85*, 78–90.

31.   Liew, Y.-M.; Heah, C.-Y.; Mustafa, A.B.M.; Kamarudin, H. Structure and properties of clay-based geopolymer cements: A review. *Prog. Mater. Sci.* 2016, *83*, 595–629. https://doi.org/10.1016/j.pmatsci.2016.08.002

32.   Khale, D.; Chaudhary, R. Mechanism of geopolymerization and factors influencing its development: A review. *J. Mater. Sci.* 2007, *42*, 729–746. https://doi.org/10.1007/s10853-006-0401-4

33.   Xu, H.; Van Deventer, J.S.J. Microstructural characterisation of geopolymers synthesised from kaolinite/stilbite mixtures using XRD, MAS-NMR, SEM/EDX, TEM/EDX, and HREM. *Cem. Concr. Res.* 2002, *32*, 1705–1716. https://doi.org/10.1016/S0008-8846(02)00859-1

34.   Snellings, R.; Mertens, G.; Elsen, J. Supplementary cementitious materials. *Rev. Mineral. Geochemistry* 2012, *74*, 211–278.

35.   Li, Z.; Ohnuki, T.; Ikeda, K. Development of Paper Sludge Ash-Based Geopolymer and Application to Treatment of Hazardous Water Contaminated with Radioisotopes. *Materials (Basel).* 2016, *9*, 633. https://doi.org/10.3390/ma9080633

36.   Nergis, D.D.B.; Abdullah, M.M.A.B.; Vizureanu, P.; Tahir, M.F.M. Geopolymers and Their Uses: Review. *IOP Conf. Ser. Mater. Sci. Eng.* 2018, *374*, 12019. https://doi.org/10.1088/1757-899x/374/1/012019.

37.   Perera, D.S.; Uchida, O.; Vance, E.R.; Finnie, K.S. Influence of curing schedule on the integrity of geopolymers. *J. Mater. Sci.* 2007, *42*, 3099–3106. https://doi.org/10.1007/s10853-006-0533-6

38.   Ghugal, Y.M. Effect of Fineness of Fly Ash on Flow and Compressive Strength of Geopolymer Concrete Article in Indian Concrete Journal. *Indian Concr. J.* 2013, *87*, 57–61.

39.   Chindaprasirt, P.; Chareerat, T.; Hatanaka, S.; Cao, T. High-Strength Geopolymer Using Fine High-Calcium Fly Ash. *J. Mater. Civ. Eng.* 2011, *23*, 264–270. https://doi.org/10.1061/(asce)mt.1943-5533.0000161.

40.   Sieve Shaker AS 200 control - RETSCH - precise sieve analysis Available online: https://www.retsch.com/products/sieving/sieve-shakers/as-200-control/function-features/ (accessed on Feb 12, 2020).

41.   Nergis, D.D.B.; Al Bakri Abdullah, M.M.; Sandu, A.V.; Vizureanu, P. XRD and TG-DTA study of new alkali activated materials based on fly ash with sand and glass powder. *Materials (Basel).* 2020, *13*. https://doi.org/10.3390/ma13020343

42.   Ahmari, S.; Zhang, L. The properties and durability of alkali-activated masonry units. In *Handbook of Alkali-Activated Cements, Mortars and Concretes*; Elsevier Inc.,

2015; pp. 643–660 ISBN 9781782422884.

43.   SO06401000 Sodium silicate, neutral solution, Scharlab S.L. The Lab Sourcing Group Available online: http://scharlab.com/catalogoproductos (accessed on Feb 12, 2020).

44.   Assaedi, H.; Shaikh, F.U.A.; Low, I.M. Influence of mixing methods of nano silica on the microstructural and mechanical properties of flax fabric reinforced geopolymer composites. *Constr. Build. Mater.* 2016, *123*, 541–552, https://doi.org/10.1016/j.conbuildmat.2016.07.049.

45.   Trejo, D.; Chen, J. *Effects of Extended Discharge Time and Revolution Counts for Ready-mixed Concrete*; 2014.

5. Chemical and Structural Analysis of Geopolymers

5.1. Chemical analysis using EDAX

The chemical composition of the obtained geopolymers shows high differences between the concentration of the main elements (Si, Na, Fe and Al). Also, all the samples contain Si, Na, Fe, Al, Ca, K, Ti, Mg and O (Figure 5.1, Figure 5.2, Figure 5.3, Figure 5.4). Except for oxygen, the chemical element with the highest concentration in all samples is silicon.

The chemical composition of sample 100FA (Table 5.1) contains the highest concentration of aluminum (Al) and iron (Fe), but the lowest concentration of silicon (Si). However, the sample 70FA_30PG exhibit a decrease of more than 20% in the concentration of Fe and Al, but due to the replacement of 30% of coal ash with glass powder, the calcium concentration in the sample increases up to 30%.

Table 5.1. Chemical composition of 100FA sample.

Element	Si	Na	Fe	Al	Ca	K	Ti	Mg	O
[%], wt.	21.46	10.98	3.14	8.86	2.25	1.05	0.20	0.73	balance
St. error, [%]	0.33	0.79	0.27	0.70	0.49	1.10	1.83	1.72	-

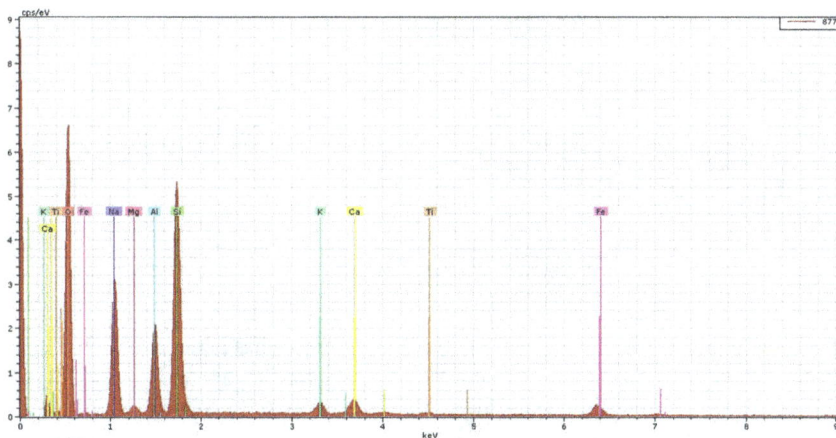

Figure 5.1. EDX spectra of the 100FA sample.

The chemical composition of the 30FA_70S sample shows an increase of approximately 17% in the Si concentration, but by replacing 70% of the amount of coal ash with sand particles, the Al and Ca content decreases by 46%, respectively 3%. In the case of 15FA_15PG_70S sample, the increase of Si concentration is of approximately 17% while the decrease of aluminum content reaches 50%. Therefore, by decreasing the coal ash content from the composition of the geopolymers, the aluminum content decrease while the silicon content increases.

Table 5.2. Chemical composition of 70FA_30PG sample.

Element	Si	Na	Fe	Al	Ca	K	Ti	Mg	O
[%], wt.	22.06	12.27	2.47	6.83	3.24	0.99	0.27	1.00	balance
St. error, [%]	0.34	0.74	0.28	0.75	0.52	1.14	1.89	1.91	-

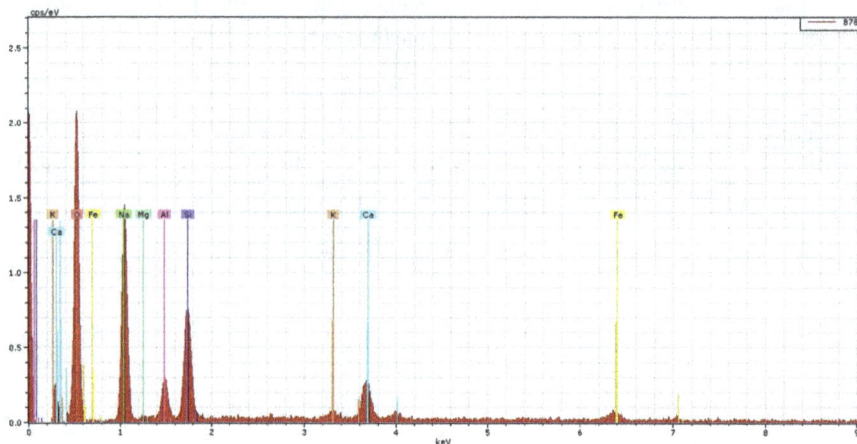

Figure 5.2. EDX spectra of the 70FA_30PG sample.

According to the chemical composition, the 100FA sample has better adhesion properties due to the ratio between the Si and Al concentration of approximately 2.4 which corresponds to a geopolymer with linear 2D structure [27,32]. However, by increasing the percentage of reinforcing particles, the Si content increases, therefore, those sample will exhibit lower flexibility but higher hardness and compressive strength.

Table 5.3. Chemical composition of 30FA_70S sample.

Element	Si	Na	Fe	Al	Ca	K	Ti	Mg	O
[%], wt.	25.72	10.83	3.08	4.79	2.20	1.13	0.29	0.80	balance
St. error, [%]	0.36	0.65	0.26	0.80	0.72	1.08	1.72	1.81	-

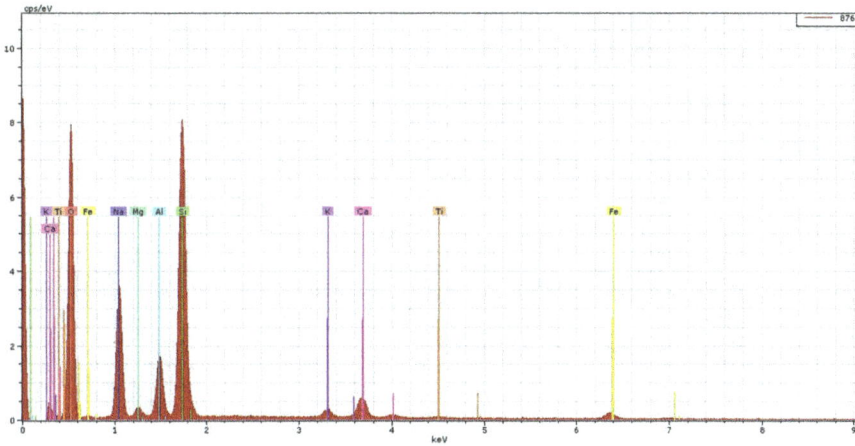

Figure 5.3. EDX spectra of the 30FA_70S sample.

The quality of the final structure and its properties results mainly from the uniformity and strength of the tetrahedral structure formed by the Si-O-Al system [9]. Any impurity or chemical element (except Si, O and Al) from the raw material, especially the calcium content, can lead to the formation of defects in the structure that significantly influence the final properties of the geopolymers [10]. To evaluate the influence of calcium content on the setting time of geopolymers, P. Chindaprasirt et al. [11] replaced different percentages of ash with Portland cement type I, which contains a high content of calcium hydroxide ($Ca(OH)_2$) and calcium oxide (CaO). According to the study, the increase in calcium content in the composition causes a sharp decrease in setting time regardless of the percentage of replacement.

5.2. Structural analysis using optical and scanning electron microscopy

The micromorphology of the obtained geopolymers has been analysed after their surface was coated with a graphite layer, which aims to reduce the background noise and thus obtain a high quality SEM micrograph. The dimensions of the studied samples differ depending on the microstructural characteristics analyzed. When the cracking mechanism was analysed, chips resulting from the destruction of the samples by compression or flexural were used as samples. When other characteristics such as the adhesion/bonding between the components, the pores distribution and geometry, the mixture homogeneity etc. cubic samples with dimensions of 10x10x10 mm³ obtained by cutting parallelepiped samples with dimensions of 20x10x10 mm³ in the direction of a plane parallel to the pouring direction passing half the length of the sample were used (see Figure 2.4).

Table 5.4. Chemica composition of 15FA_15PG_70S sample.

Element	Si	Na	Fe	Al	Ca	K	Ti	Mg	O
[%], wt.	26.41	11.01	2.30	4.42	2.83	0.81	0.73	0.90	balance
St. error, [%]	0.35	0.69	0.27	0.77	0.62	1.11	1.79	1.72	-

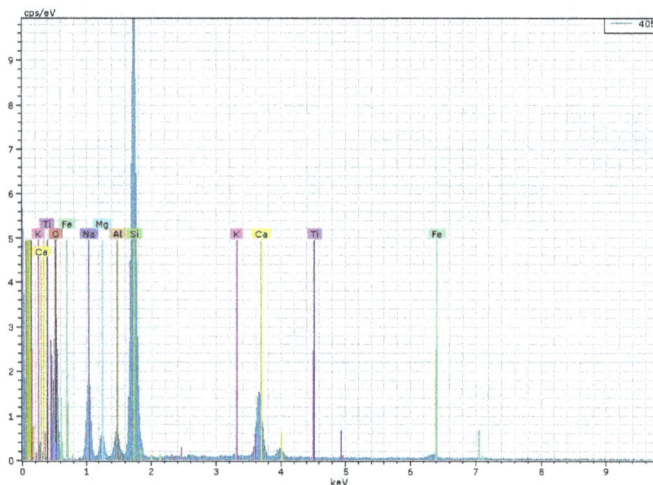

Figure 5.4. EDX spectra of the 15FA_15PG_70S sample.

The main advantage of microstructural examination by optical microscopy is the highlighting of the multicomponent characteristic of geopolymers, based on the differences in color and brightness of the compounds. However, at high magnification the surface topography can only be obtained by electron microscopy (SEM), because of the irregular geometric shape of the compounds which leads to a high decrease in light beam reflection, therefore, a poor quality image is obtained.

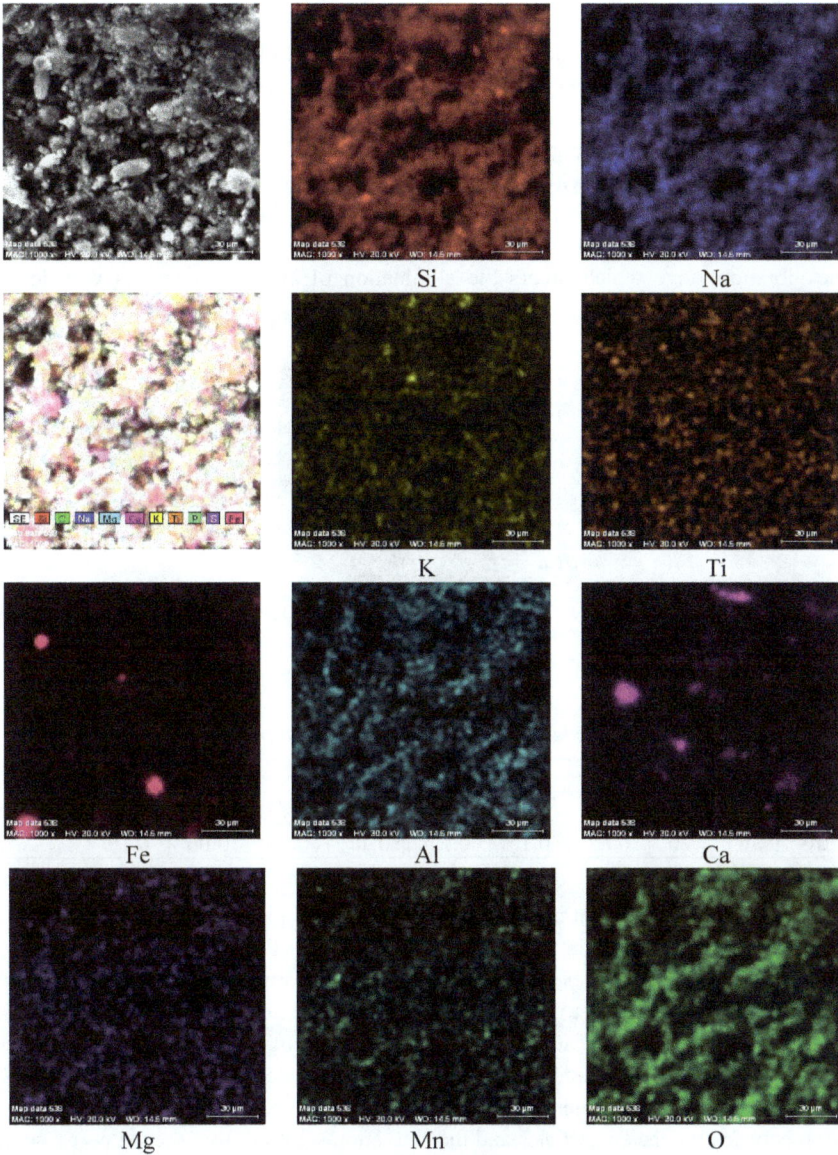

Figure 5.5. Elemental mapping of 100FA sample.

Another advantage of using the scanning electron microscope is the possibility of simultaneous analysis of the microstructure and elemental chemical distribution on the surface of the samples. From a chemical-structural point of view, the samples show a homogeneous elemental chemical distribution (Figure 5.5). However, from a macroscopic point of view (Figure 5.6) the formation of a layer in the upper part of different color (Figure 5.7.a) with a thickness of about 3 mm can be observed. This layer shows a pattern specific to the elimination of air bubbles and contains unreacted particles and cracks in large numbers (Figure 5.7.b). The formation of the upper layer may be due to the vibration stage which favors the elimination of air and impurities with lower density from the mixture, during the binder phase.

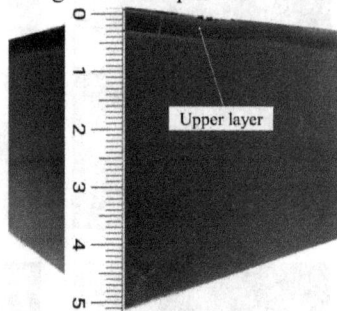

Figure 5.6. Geopolymers samples morphology.

Table 5.5. Chemical composition of the upper layer.

Element	Si	Na	Fe	Al	Ca	K	Ti	Mg	O
[%], wt.	27.80	12.09	4.82	11.46	2.77	2.77	0.97	1.17	balance
St. error, [%]	1.20	0.82	0.15	0.57	0.11	0.11	0.06	0.10	-

From the microstructural point of view, the samples show a homogeneous structure with high compactness of the matrix that depends on the type or the percentage of reinforcing particles. Moreover, all samples show unreacted coal ash particles, but the sample 100FA (Figure 5.8) shows the highest number, therefore, in some areas the matrix continuity is interrupted. Additionally, the SEM micrograph at 500X also highlights the cracks formed during the curing process, due to the fast evaporation of water. However, due to the self-healing capacity of geopolymers, which is related to the continued geopolymerisation reaction between unreacted particles and the activation solution from the gel pores, some cracks are repaired over time. At the same magnification, the sample 70FA_30PG (Figure 5.9) shows a homogeneous structure, with a high degree of ash dissolution.

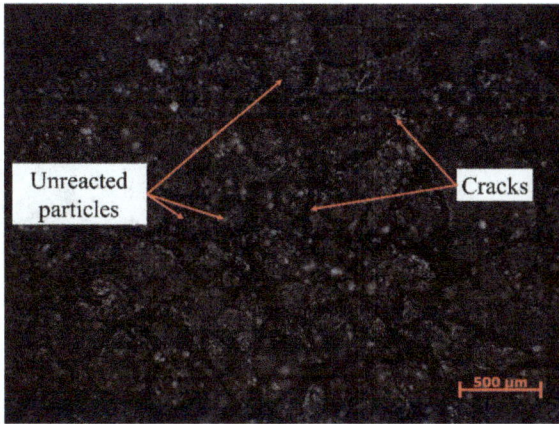

a)

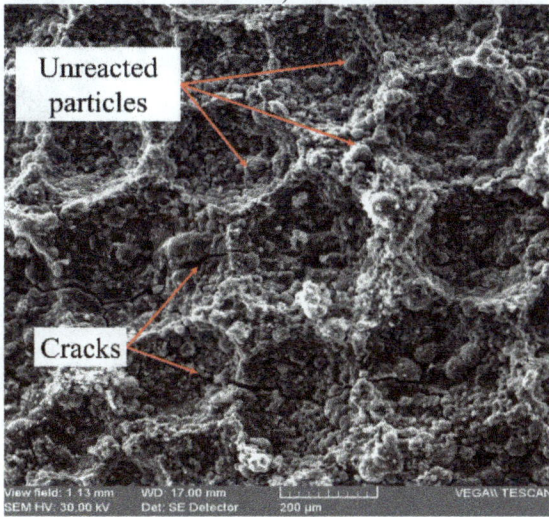

b)

Figura 5.7. Morfologia stratului exterior. a) micrografie optică;
b) micrografie SEM.

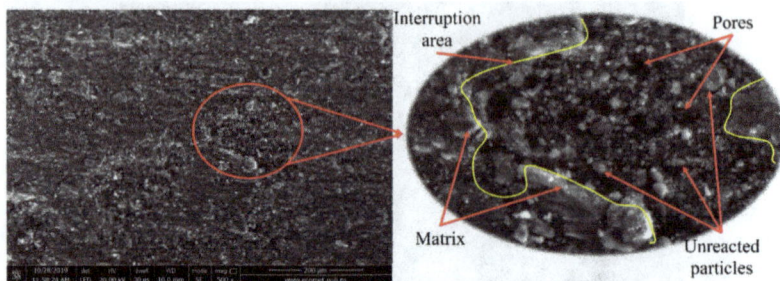

Figure 5.8. The micromorphology of 100FA sample.

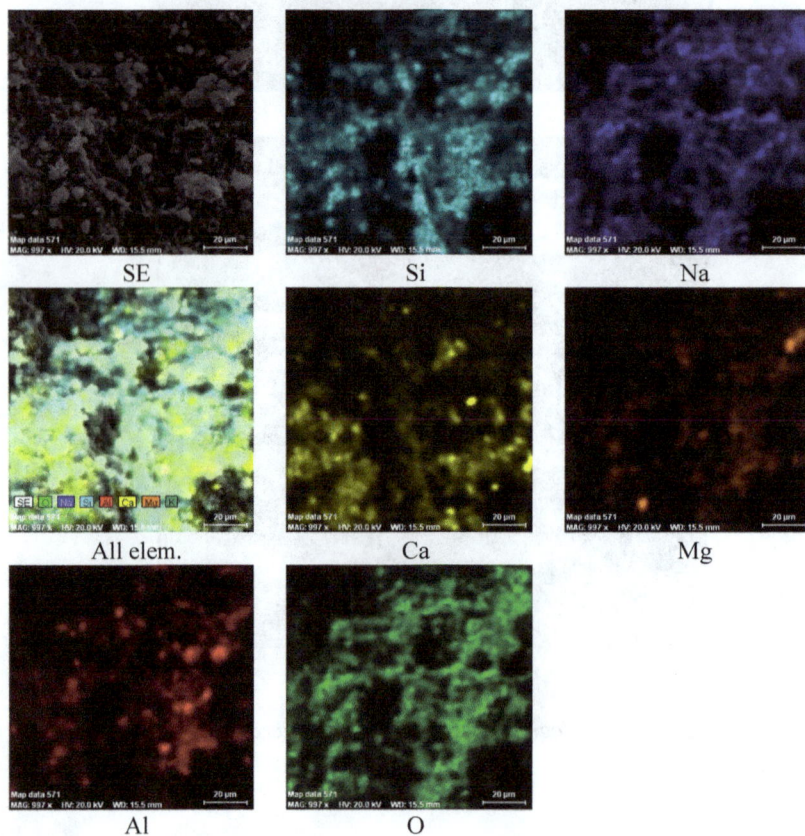

Figure 5.9. Elemental mapping of 70FA_30PG sample.

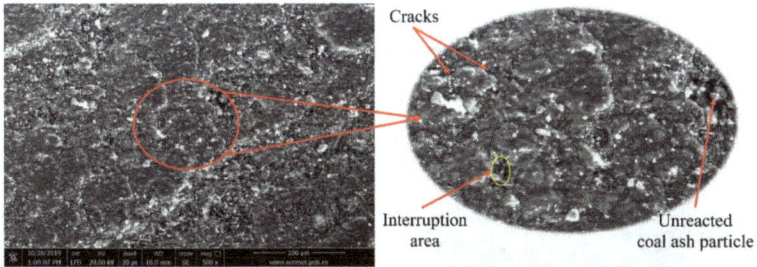

Figure 5.10. The micromorphology of 70FA_30PG sample.

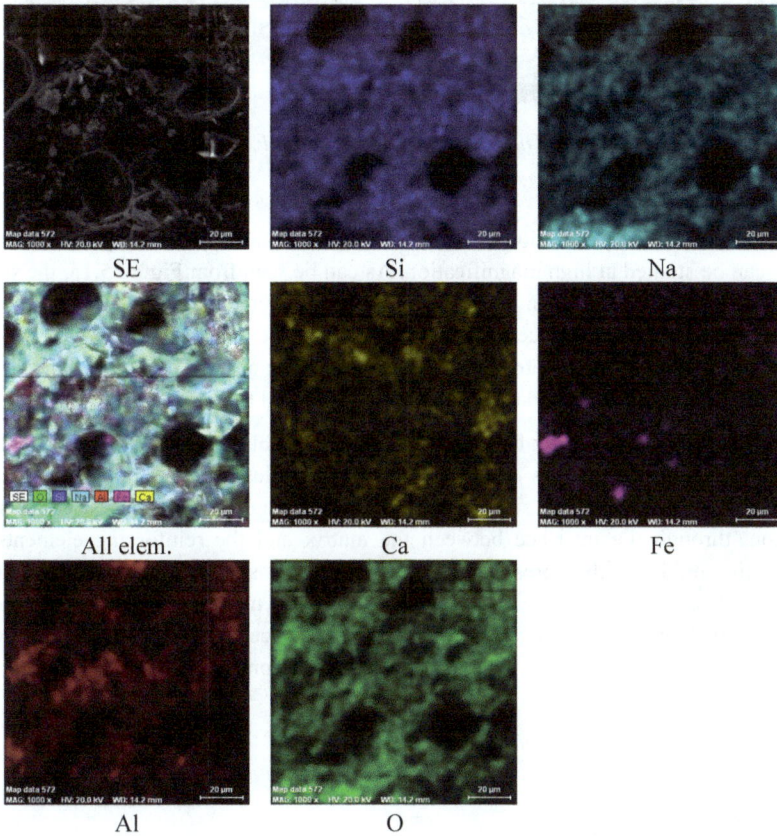

Figure 5.11. Elemental mapping of 30FA_70S sample.

After sand particles were introduces into the structure, the number of large pores increases, but there is a significant decrease in the number of unreacted particles and the number of small pores (Figure 5.12 and Figure 5.14).

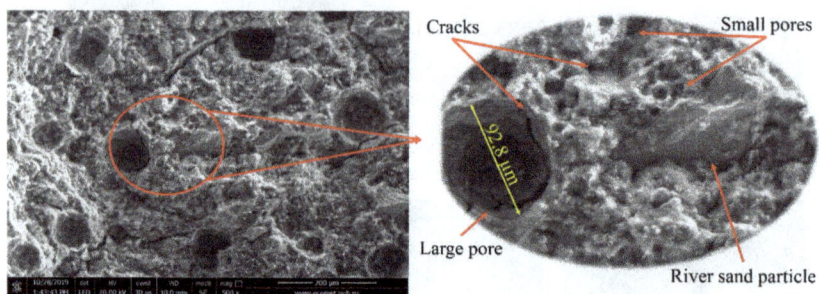

Figure 5.12. The micromorphology of 30FA_70S sample.

In environmental SEM condition, the interface between the reinforcing particles and the matrix can be studied at high magnification. As can be seen from Figure 5.15, the surface of the glass particles reacted in the alkaline medium (reacted zone), therefore, the matrix adheres strongly to their surface. In the case of the samples with river sand as reinforcing elements (Figure 5.16) the matrix does not adhere to the surface of the particles, thus, a clear delimitation can be observed between the matrix and the sand particles.

The microstructural analysis of the fracture surface highlights the cracking mechanism during the compression stress of the geopolymer samples. The 3D optical micrograph (Figure 5.17) of the breaking surface confirms the sliding of the cracks, in a sinusoidal direction, through the interface between the matrix and the reinforcing elements and through the middle of the pores. Thus, the inner pores, especially the large ones, act as structural defects that produce negative effects on the main mechanical properties of geopolymers. However, the negative effects they introduce, in the case of samples with a high percentage of large pores, are offset by the high compressive strength of the sand particles.

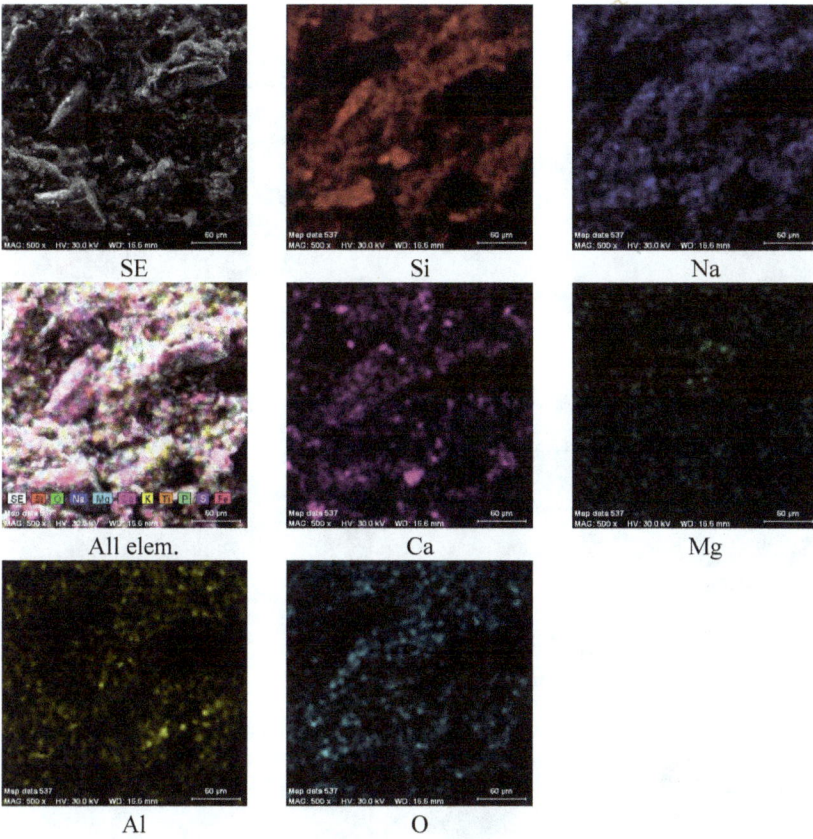

Figure 5.13. Elemental mapping of 15FA_15PG_70S sample.

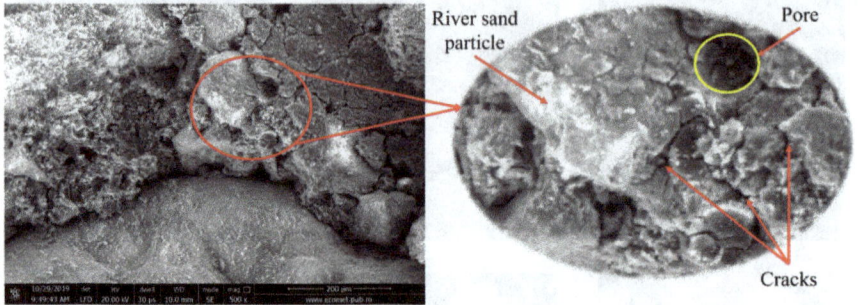

Figure 5.14. The micromorphology of 15FA_15PG_15S sample.

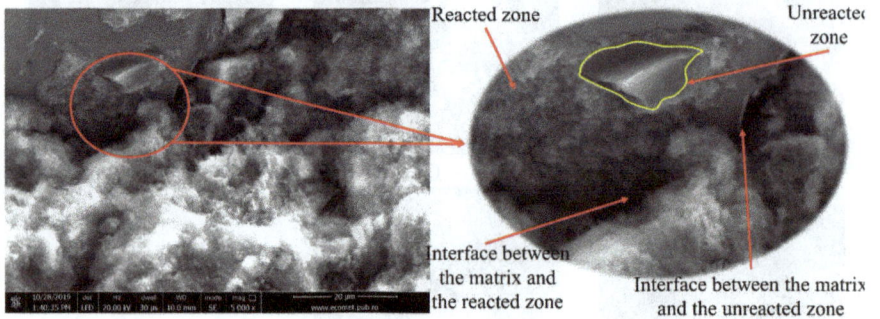

Figure 5.15. 70CA_30GP sample micromorphology at high magnification (10kX).

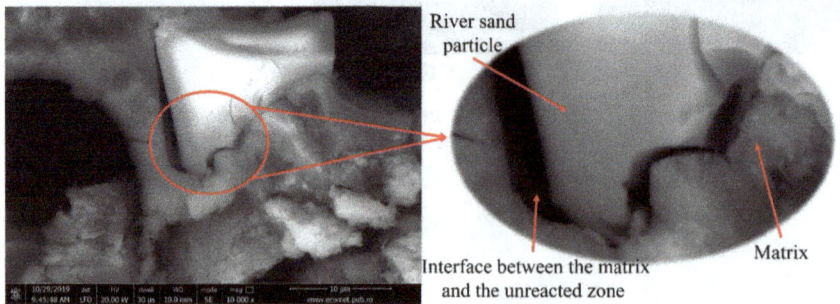

Figure 5.16. 30CA_70RS sample micromorphology at high magnification (10kX).

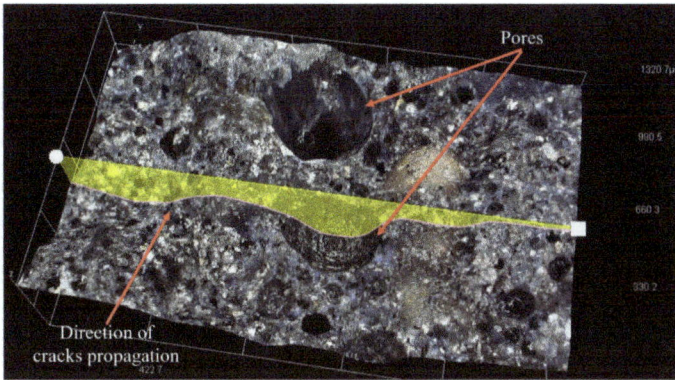

Figure 5.17. Micrografie optică 3D a suprafeței de rupere a probei 15FA_15PG_70S.

Analyzing the matrix of the geopolymer at high magnification (Figure 5.18) we can highlight the spongy (porous) structure formed by three categories of pores. From a dimensional point of view, the pores present in the matrix of geopolymers based on power plant ash are divided into: gel-type pores with a diameter of less than 50 nm, capillary pores with a diameter between 50 nm and 600 nm and pores with large diameters (over 600 nm) [3].

The distribution and size of the pores can adversely affect, in particular, the main mechanical properties of geopolymers, as it reduces the compactness of the structure and allows water or acidic substances to enter the sample. In the study of Farhana F. [5] it was found that total porosity and pore size distribution are the most important factors affecting the compressive strength.

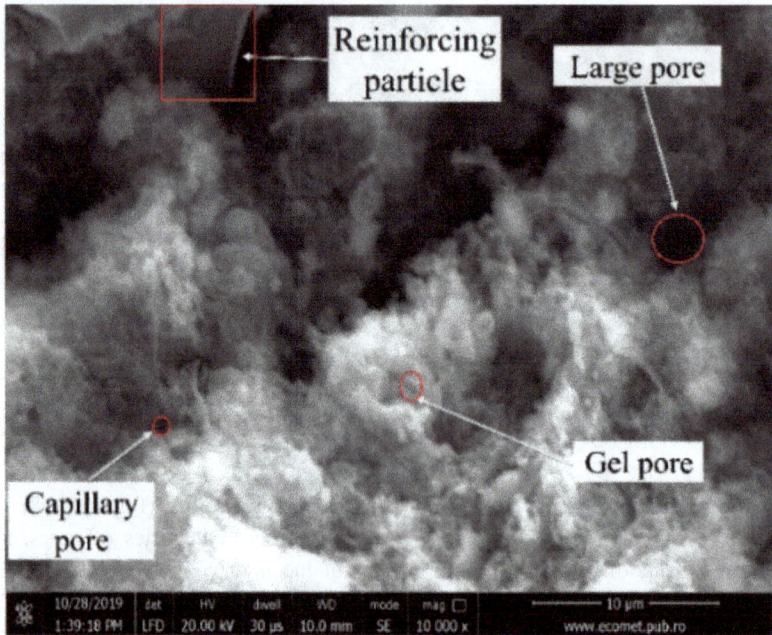

Figure 5.18. Coal-ash based geopolymers morphology [4].

5.3. Mineralogical analysis by X-ray diffraction

The determination of the mineralogical changes produced by the alkaline activator on the power plant ash was performed on $10\text{x}10\text{x}10 \text{ mm}^3$ geopolymer samples maintained under normal atmospheric conditions for 90 days and ground 24 hours before analysis.

The change of the initial phases, specific to the raw material, in other phases, specific to zeolites (materials with a honeycomb cubic structure with Al and Si atoms surrounded by 4 oxygen atoms) is governed by the characteristics of the geopolymerization reaction. This chemical reaction between the alkaline solution and the gray compounds of the power plant can be divided into two stages. In the first stage the raw material is dissolved under the action of the alkaline solution forming reactive precursors of $Si(OH)_4$ and $Al(OH)_4$. This step is crucial for the geopolymerization reaction because it can limit the rate of structure formation. In the second phase, the polymerization and precipitation of the system takes place, resulting in the condensation of Si-O-Al molecules into various compounds.

The diffractogram specific to the coal ash analysis (Figure 5.19) shows mainly peaks specific to the complex chemical compounds formed between the main component chemical elements, such as quartz (Θ), corundum (M), anorthite (A), mullite (γ) or hematite (H), but also other secondary phases, from a quantitative point of view, including calcium, titanium, iron or magnesium, such as Γ - goethite, 1 - bleached, calcium carbonate (Xα), calcium hydroxide (Aλ), goethite (Γ), aluminum hydroxide (Π) etc. Both in the case of the diffractogram of the raw material and in the case of the one specific to the geopolymers obtained, most of the peaks points are positioned between 20° and 45° (2θ). At the same time, the peak with the highest intensities, specific to quartz and corundum, are positioned between 25° and 30° (2θ). The diffractogram specific to glass powder (Figure 5.12) shows a model specific to amorphous materials.

Due to the fact that these types of geopolymers consist of several phases that have similar XRD models. If a peak corresponds to several phases, only the phase with the highest intensity was presented on the model in the figures. Therefore, some peaks initially assigned to some phases, after the TG-DTA analysis, were assigned to other phases because the phase intensity is different, e.g. the maximum intensity at 21.09 °, 2θ obtained before the TG-DTA analysis corresponds to goethite, but after the TG-DTA analysis the peak positioned at 20.85°, 2θ corresponds to quartz.

The quartz or silicon dioxide detected is a mineral with a tetrahedral structure formed between silicon atoms and oxygen that crystallizes in the hexagonal system. Its concentration positively influences the mechanical properties of geopolymers, because quartz particles form barriers that oppose the propagation of cracks.

The detected corundum crystallizes in the rhombohedral system being known as one of the main oxides of aluminum. This compound is essential for geopolymers because its hardness is close to that of diamond.

Hematite crystallizes in the rhombohedral system being a compound of iron with oxygen. It has the same crystallographic structure as that of corundum, being frequently found next to it.

Goethite crystallizes in the orthorhombic system being an acicular iron mineral composed of oxygen and hydrogen. It has low hardness and instability at high temperatures.

Anorthite crystallizes in the triclinic (anortic) crystalline system being the compound with the richest calcium content in the group of plagioclase feldspars. It is found in several colors and consists of calcium, aluminum, silicon, oxygen, but also potassium, sodium, iron and titanium at the trace level.

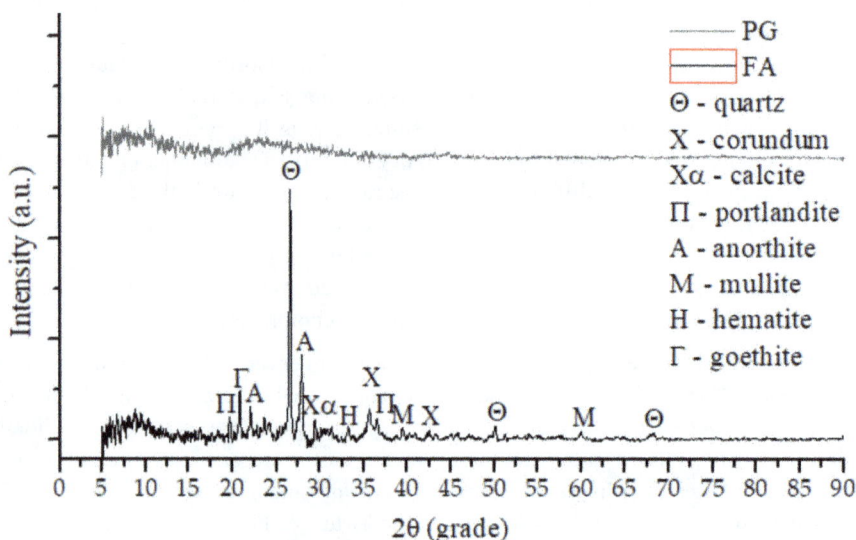

Figure 5.19. Mineralogical characterization by X-ray diffraction. a) power plant ash (black); b) glass powder (grey).

Sodalite crystallizes in the cubic system being a complex mineral formed by the reaction between sodium or chlorine with the main elements of the raw material (aluminum, silicon and oxygen). Natural sodalite consists of an Al-O-Si network that incorporates Cl^+ cations, but the one resulting from geopolymerization has interstructural Na^+ cations, similar to zeolites [6].

Following the chemical geopolymerization reaction between the ash of the thermal power plant and the activation solution of the main phase specific to the raw material, quartz, whose peak is positioned at 26.62°, 2θ decreases in intensity due to the decrease of the glass phase, but there is a significant increase in intensity. anorthite, 27.94°, 2θ and the appearance of new phase-specific peak created by the reaction between Na^+ and the other compounds (Figure 5.20).

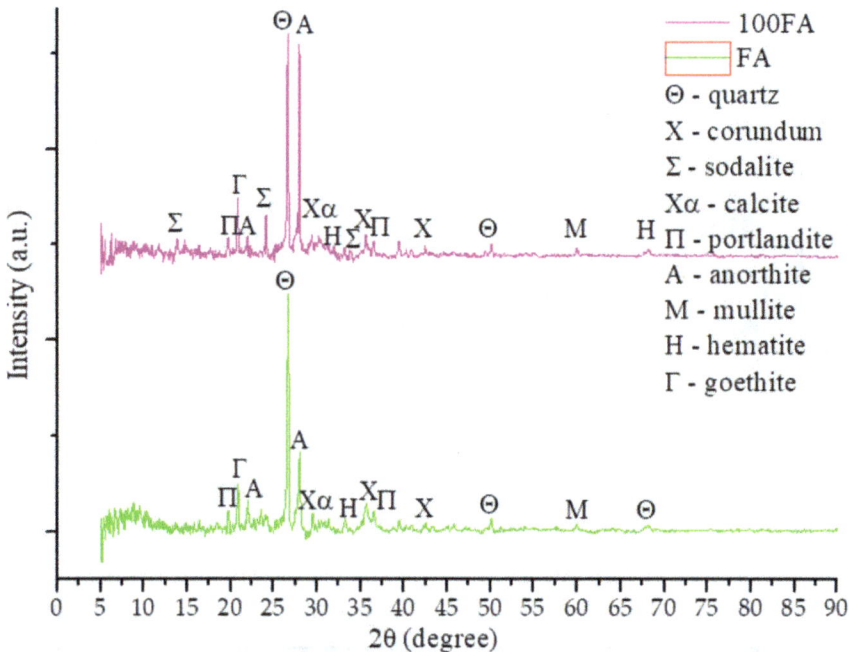

Figure 5.20. Mineralogical characterization by X-ray diffraction. a) power plant ash (green); b) sample 100FA (magenta).

The diffractogram specific to sample 100FA (Figure 5.20) shows the formation of the most important phase specific to geopolymerization, sodalite (S), it shows three peaks between 8° and 35°, 2θ. The appearance of such a phase specific to zeolites suggests the formation of a mesoporous material (pores with a diameter between 20 and 50 nm) of semi-crystalline nature [7–10]. The sodalite content formed being directly proportional to the cation exchange capacity between the raw material and the activation solution [11,12].

By replacing the ash of the power plant with glass powder, there are changes on the diffractogram characteristic of sample 100FA which consist in decreasing the peak intensity specific phases that highlight the chemical reaction between the activation solution and ash and increasing the intensity of specific quartz (Figure 5.21). Following the replacement of 70% of the power plant ash specific to the 100FA sample with sand particles, it was found that the intensity of some phases increased greatly (Figure 5.22 - quartz and anorthite). At the same time, the phase-specific peaks, which highlight the geopolymerization, decrease significantly as a result of the reduction of the Al content available in the system (Figure 5.21 - sodalite).

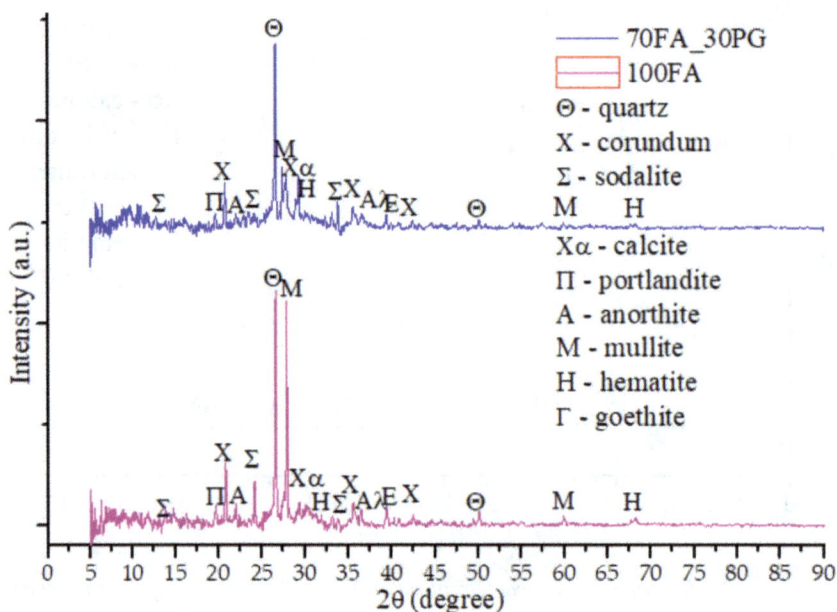

Figure 5.21. Mineralogical characterization by X-ray diffraction. a) power plant ash (magenta); b) sample 70FA_30PG (blue).

The XRD diffractogram of sample 15FA_15PG_70S (Figure 5.23) shows a decrease in intensity of the peak corresponding to the compounds with aluminum content, due to the reduction of the ash content of the thermal power plant in the sample.

To confirm the decomposition of calcium hydroxide and goethite during heating, XRD analysis was also performed on the samples studied by TG-DTA. Therefore, diffractograms specific to samples heated to 1000 °C show changes in peak intensity (Figure 5.24), especially for calcite and hematite, due to chemical reactions [4] of decomposition of calcium hydroxide and goethite and formation of carbonate. calcium and hematite. Moreover, because the geopolymerization continues during heating, the anorthite reacts with Na^+ cations and forms a new, bleached (ε) phase, the peak of which is positioned at 27.85°, 2θ.

The presence of clear diffraction peaks is an indication of consistent long-range ordering. Only the diffraction peaks of the perfect crystals are very narrow. This is a limitation of the method, but also an advantage: The width of the diffraction peaks provides information about the dimensions of the reflecting planes.

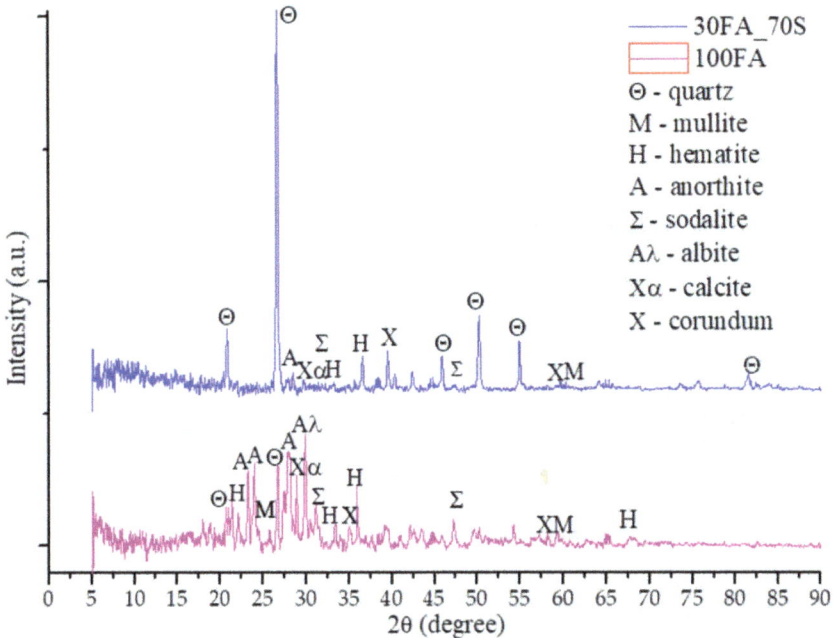

Figure 5.22. Mineralogical characterization by X-ray diffraction. a) power plant ash (magenta); b) sample 30FA_70S (blue).

Although, from a phasic point of view, an increase in the calcium oxide content can be observed in the glass samples, which positively influences the compressive strength of the concrete. In old age, it has been shown that the compressive strength of geopolymers with natural aggregates is higher. However, from the results presented it can be seen that glass can become a successful replacement for natural aggregates, especially in materials that use alkaline activators. Their fineness gives a high increase in the density of the samples, thus supporting their abrasion resistance.

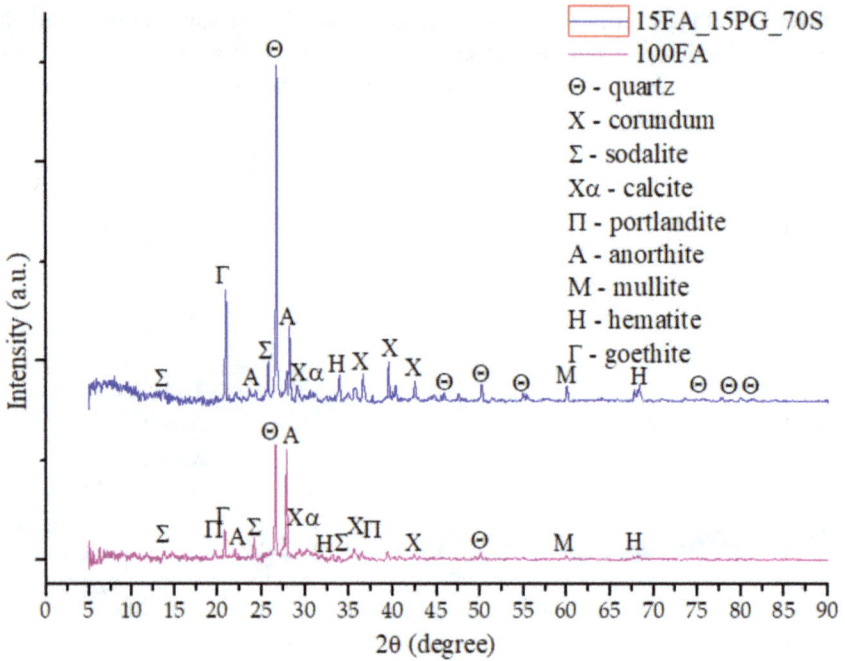

Figure 5.23. Mineralogical characterization by X-ray diffraction. a) power plant ash (magenta); b) sample 15FA_15PG_70S (blue).

The sand used is the fine part of the cane used in civil engineering applications. Its extraction / separation is done after washing and sorting the ballast, in order to remove organic impurities. The control screens used to determine the granularity of the natural aggregates have square meshes. The sieve set must include, in any case, depending on the dimensions of the product, the following nominal dimensions: 0.063 mm, 0.125 mm, 0.250 mm, 0.500 mm, 1 mm, 4 mm, 8 mm, 16 mm, 25 mm, 31.5 mm, 40 mm, 63 mm. The nature and petrographic-mineralogical characteristics were verified according to the standards in force. It is required that the approval of ballast products and periodic checks, examination of aggregates, be performed by a qualified geologist. Natural aggregates must not contain foreign bodies, pyrite, limonites or soluble salts. It is forbidden to use natural aggregates with a content of granules consisting of altered, soft, friable, porous rocks. Thus, a minimum content of impurities was ensured in the geopolymer samples with reinforcing particles.

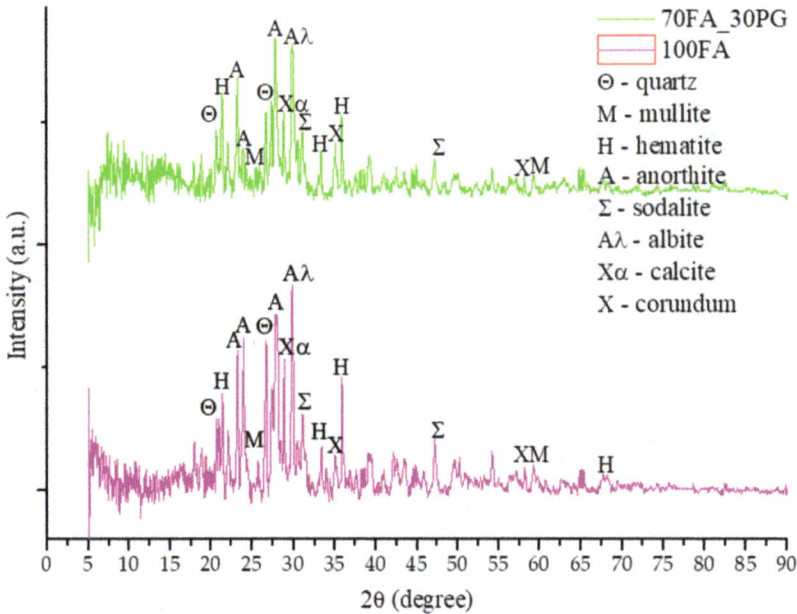

Figure 5.24. Mineralogical characterization by X-ray diffraction of high-temperature exposed samples (after TG-DTA analysis). a) sample 100FA (magenta); b) sample 70FA_30PG (green).

Most of the OH⁻ groups related peaks disappear after the samples were heated up to 900 °C. The various degrees of dehydroxylation of goethite or portlandite were attributed to structural and chemical rearrangements that caused decomposition of the tetrahedral and octahedral compounds. This caused displacement of the diffraction peaks to lower angles and an inversion of the relationship between the intensities of the compounds with chemical water bonded molecules peaks after calcination. Goethite was transformed into hematite in the calcinated geopolymers, this behaviour was also observed in (Azeredo Melo, 2007) on metakaoline based geopolymers.

Table 5.6. Peaks assigning from geopolymer-specific diffractograms.

Phase	d, [A]	2θ	I [%]	Phase	d, [A]	2θ	I, [%]	Phase	d, [A]	2θ	I, [%]
Σ	6.30	14.05	17	1	2.56	35.02	12	Aλ	1.67	54.94	21
Aλ	4.68	18.95	48	Σ	2.56	35.02	13	Χα	1.62	56.78	40
A	4.44	20.00	13	M	2.53	35.47	100	H	1.60	57.51	8
Θ	4.26	20.85	22	H	2.52	35.61	70	Χα	1.6	57.55	50
Γ	4.21	21.09	100	A	2.52	35.66	13	Γ	1.60	57.56	8
1	4.05	21.92	35	Χα	2.49	36.04	60	M	1.59	58.05	93
A	4.04	22.01	13	Aλ	2.46	36.49	100	Aλ	1.57	58.65	9
Χα	3.84	23.14	60	Θ	2.46	36.53	7	Γ	1.56	59.18	28
1	3.8	23.39	16	Γ	2.45	36.65	80	Σ	1.56	59.18	10
A	3.74	23.78	72	1	2.44	36.80	14	Θ	1.54	59.93	6
H	3.69	24.13	33	Σ	2.37	37.93	17	Χα	1.52	60.89	60
Π	3.66	24.29	25	M	2.36	38.09	47	Σ	1.52	60.90	5
A	3.64	24.42	13	Π	2.32	38.78	12	Γ	1.50	61.80	24
A	3.64	24.42	13	Χα	2.28	39.49	70	M	1.50	61.93	7
Σ	3.63	24.50	100	Γ	2.25	40.04	12	H	1.49	62.39	22
M	3.45	25.79	69	H	2.21	40.84	17	Σ	1.48	62.73	7
Γ	3.39	26.27	12	Γ	2.19	41.19	20	Χα	1.47	63.20	40
A	3.36	26.54	26	Χα	2.09	43.25	70	Γ	1.46	63.69	12
Θ	3.35	26.62	100	Σ	2.08	43.47	33	H	1.45	63.97	21
A	3.21	27.73	100	M	2.07	43.73	96	Χα	1.44	64.67	50
A	3.2	27.85	100	Χα	1.92	47.30	90	Σ	1.43	65.19	8
Χα	3.02	29.55	100	Γ	1.92	47.31	8	M	1.39	67.14	35
Π	2.96	30.16	25	Χα	1.87	48.65	80	Aλ	1.38	67.76	9
A	2.95	30.29	10	H	1.84	49.42	31	Aλ	1.37	68.26	6
Aλ	2.89	30.89	19	Aλ	1.82	50.10	38	M	1.36	68.83	54
Σ	2.81	31.82	5	Θ	1.82	50.11	10	Γ	1.36	69.00	8
H	2.70	33.12	100	Γ	1.80	50.67	8	Γ	1.32	71.40	12
Γ	2.70	33.15	36	M	1.73	53.03	47	H	1.31	71.82	7
Γ	2.58	34.74	24	Γ	1.72	53.21	36	Aλ	1.23	77.54	6
A	2.57	34.94	20	H	1.70	54.01	36	M	1.23	77.72	15

From a mineralogical point of view, the geopolymerization reaction produces the transformation of rich compounds into silicon and aluminum, especially quartz and anorthite, into a compound specific to zeolites (sodalite).

5.4. Structural characterization by FTIR spectroscopy

Fourier transform IR spectroscopy is a non-destructive technique for analyzing the chemical structure of a material. This consists of obtaining an infrared light absorption spectrum at different wavelengths of a beam. The analysis was performed using a Bruker Hyperion 1000 FTIR spectrometer, coupled with a microscope, equipped with a 15X lens. Due to the use of the microscope, the samples didn't need to be embedded in KBr pellets. Therefore, the analysis was carried out directly on polished samples in a range of wave numbers between 4000 cm^{-1} and 600 cm^{-1} with a resolution of 4 cm^{-1} at a scan frequency of 10 kHz through a 6 mm diameter aperture and 64 scans for each surface. The absorbance spectrum of the samples show multiple peaks included in the vibration bands of the chemical bonds in the present groups. The spectra were analyzed using OPUS 65 Bruker (Bruker, Germany) software to study, in particular, the groups formed between Si, Al, H, and O.

The chemical structure of the obtained geopolymers presents multiple vibration bands specific to OH$^-$ and Si groups (Si-OH), asymmetric Si-O-Si groups, asymmetric Si-O-Al groups and Si-O rings. The FTIR spectra (Figure 5.25) of coal-ash (FA) and glass powder (PG) shows a wide vibration band (I) between 3700 cm^{-1} and 3000 cm^{-1} which is attributed to the stretching vibration and bending vibration of OH$^-$ groups [38]. The large bandwidth is due to the high degree of hydrogen association with other hydroxyl groups by creating strong links between the OH$^-$ and Si ($\equiv$Si-OH) groups. The second significant peak (II), between 1150 cm^{-1} and 1250 cm^{-1}, can be associated with the specific rhythmic band along the covalent bond axis, which is known as stretching vibration of the asymmetrical Si-O-Si groups [39]. The vibration band between 800 cm^{-1} and 700 cm^{-1} (III) is specific to the asymmetric Si-O-Al groups in the compounds [40] (corundum, anorthite) of the analyzed material, and the band between 700 cm^{-1} and 600 cm^{-1} is attributed to the Si-O rings [41].

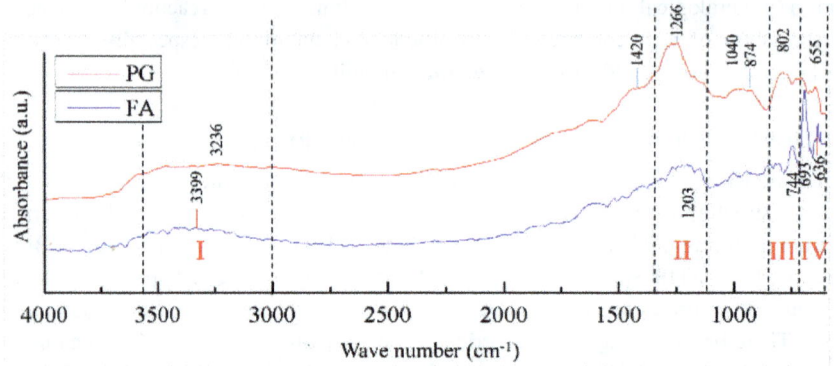

Figure 5.25. FTIR spectra of raw material (FA) and glass powder (PG).

The chemical structure analysis of the raw materials through mFTIR confirms the presence, on the analyzed surface, of hydroxide groups and silicon, oxygen and aluminum compounds which correspond to a high percentage of quartz, corundum, anorthite and vitreous phase.

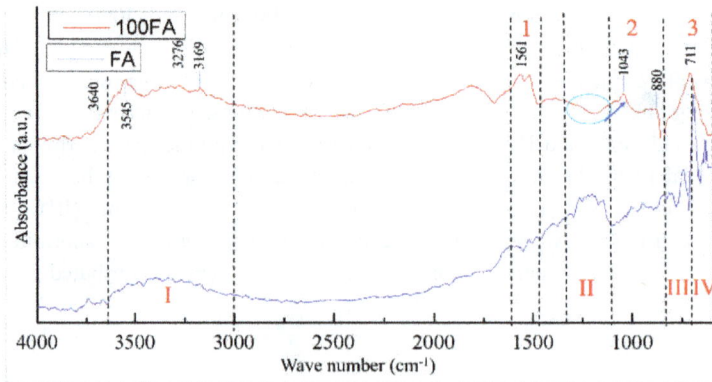

Figure 5.26. FTIR spectra of raw material (FA) and sample 100FA (100FA).

As a result of the coal-ash activation (Figure 5.26), the vibration band I show an increase in intensity due to the Si-OH bond's appearance and OH⁻ group concentration increase. Simultaneously, another vibration band (1) between 1650 cm⁻¹ and 1480 cm⁻¹ appears, which corresponds to the change of the angle between two covalent bonds, known as deformation vibration. In this case, the deformation vibration of the δ-HO-H bonds between the hydrogen and oxygen atoms specific to the adsorbed water molecules [42] is recorded. The bands III and IV undergo significant transformations as a result of the reaction between the compounds rich in aluminum and silicon and the alkaline activator, thus a specific vibration band (3) appears. The position change of the band is attributed to the internal vibrations of the sialates tetrahedra (Si-O-Al, Si-O-Al-O-Si-O or Si-O-Al-O-Si-O-Si-O) resulting from geopolymerisation [24]. Moreover, band II is also shifted to low frequencies (2) as a result of the increase in OH⁻ groups concentration on the analyzed surface and also due to Al^{3+} atoms penetration into the initial Si-O-Si structure forming the N-A-S-H and C-A-S-H phases, a phenomenon specific to zeolites [43]. A high peak corresponds to a high rate of the aluminum atom in the $[SiO_4]^{4-}$ group penetration, i.e. a higher content of N-A-S-H and C-A-S-H. The main element of influence is the sodium ions concentration in the activator that cause the Si-O bond breakage and increase the ability to incorporate aluminum during the gel phase [44]. This aspect is confirmed by the appearance of the crystalline phase in the structure and the increase of the hygroscopicity of the material by the appearance of the small pores that can be observed in the SEM micrographs (see Figure 5.12).

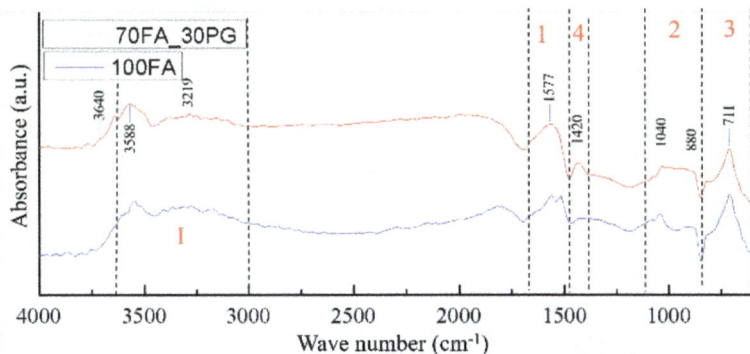

Figure 5.27. FTIR spectra of sample 100FA (100FA) and sample 70FA_30PG (70FA_30PG).

Besides the main vibration bands which correspond to silicon, oxygen, aluminum and hydrogen compounds, another peak at 3640 cm^{-1} which corresponds to OH$^-$ stretching vibration appears. Following the replacement of 30% of coal-ash with glass powder, the FTIR spectrum (Figure 5.27) doesn't present significant changes. However, the band I is higher due to the increase of hydroxide groups, introduced in the sample by the activation solutions, to raw material ratio. Moreover, a new band (4) between 1480 cm^{-1} and 1370 cm^{-1} appears, which is attributed to the C-O groups in CO_3^{2-} of calcite, especially, from glass particles [45].

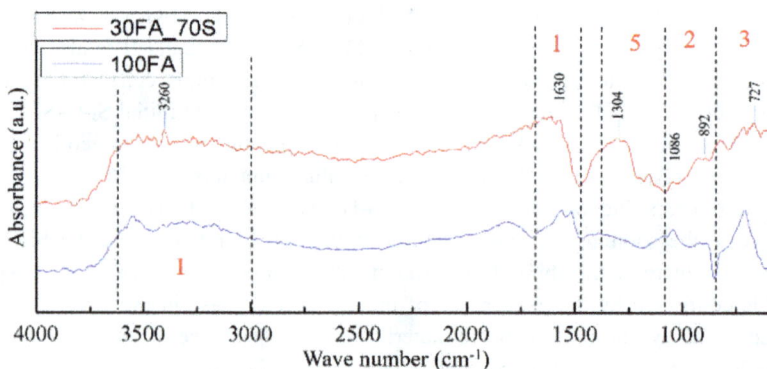

Figure 5.28. FTIR spectra of sample 100FA (100FA) and sample 30FA_70S (30FA_70S).

After replacing 70% of coal-ash with sand particles two significant peaks, between 1250 cm^{-1} and 1100 cm^{-1}, appear on the sample 30FA_70S FTIR spectrum (Figure 5.28). These peaks correspond to the stretching vibration band of the asymmetric groups of ν(Si–O–Si) and δ(Si–O) [46]. These two groups are characteristic to the quartz from the sand.

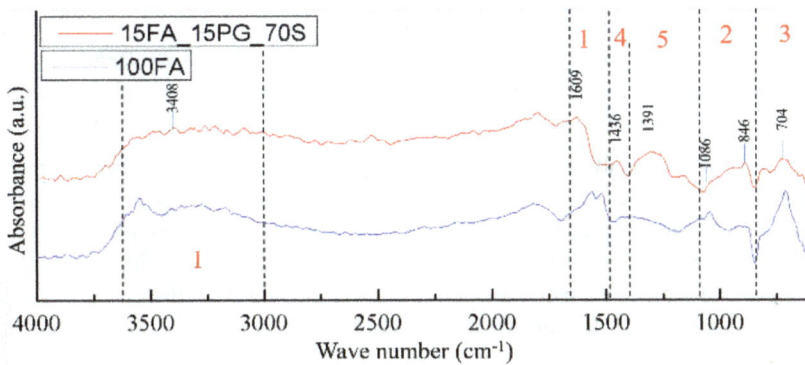

Figure 5.29. FTIR spectra of sample 100FA (100FA) and sample 15FA_15PG_70S (15FA_15PG_70S).

The sample 15FA_15PG_70S FTIR spectrum (Figure 5.29) shows a decrease in the intensity of peaks from the bands 2 and 3 as a result of the aluminum concentration decrease. Therefore, the coal-ash concentration is directly proportional to the number of sialates groups present in the structure.

FTIR spectroscopy analysis of the obtained geopolymer samples reveals their chemical structure- mainly based, on groups formed between silicon, oxygen and aluminum atoms, but also hydrogen. As a result of the activation, on the FTIR spectra also appear bands specific to the sialates that confirm the geopolymerization reaction between the raw material and the activator. Moreover, the FTIR spectra of the analysed samples show a band specific to water molecules that highlights the hygroscopic characteristics of these materials.

The analysis of the chemical composition of the samples highlights the differences produced by the reduction of the amount of power plant ash by replacing it with reinforcing particles. Sample 100FA contains the highest concentration of Si, Fe, Al, Ca, K, Ti and Mg, but the lowest concentration of Na. Sample 70FA_30PG shows a decrease of approximately 7% of the Si concentration and of more than 10% of the Fe, Al and Ca concentration as a result of the replacement of 30% of the ash of the power plant with glass powder. The chemical composition of the 30FA_70S sample shows an increase of approximately 44% in the Na concentration, and in the case of the 15FA_15PG_70S sample the increase is 48%. The phenomenon is due to a significant reduction in the amount of ash relative to the amount of activation solution.

From a structural point of view, the geopolymer samples obtained have a homogeneous chemical distribution. However, from a macroscopic point of view, a different layer with a thickness of about 3 mm is observed at the top of the samples. Its morphology suggests a separation, during the vibration stage, of impurities based on differences in density.

The microstructural analysis of the 100FA sample reveals a large number of unreacted ash particles throughout the volume, as well as cracks formed during the drying process due to the rapid evaporation of water. After the introduction of the reinforcing particles into the structure (samples 70FA_30PG, 30FA_70S and 15FA_15PG_70S), a decrease in the number of unreacted particles can be observed, however, there is a significant increase in the number of large pores (> 600 nm) due to blockage of air bubbles. by aggregates.

During the investigation of the compressive strength of the samples, the cracks formed advance through the interface between the matrix and the reinforcing elements, respecting the law of lowest resistance. The microstructural analysis of the samples highlights their cracking mechanism, noting that the propagation of cracks takes place after a sinusoidal trajectory through the interface between the matrix and the reinforcing elements, but also through the middle of the pores.

From a mineralogical point of view, the ash of the power plant used contains mainly oxide compounds of aluminum and silicon. Most of the peaks present on the diffractograms are positioned between 20° and 45° (2θ), and the phase corresponding to the peak with the highest intensity is quartz (26.62°, 2A). Following activation, an additional phase specific to zeolites appears, sodalite (24.5°, 2A), which confirms that the activation of the geopolymer took place.

The analysis by IR spectroscopy with Fourier transform of the obtained geopolymer samples highlights their chemical structure based mainly on groups formed between silicon, oxygen and aluminum atoms, but also hydrogen. Following activation, in addition to the characteristic gray strips of the power plant, glass particles or sand, there are also sialate-specific strips (Si-O-Al) that confirm the geopolymerization reaction between the raw material and the activator, as well as a specific band of water molecules, tape that highlights the hygroscopic character of these oxide materials.

The FTIR spectrum of the power plant ash and the glass powder used shows a wide vibration band between 3700 cm^{-1} and 3000 cm^{-1} which is assigned to the OH^- groups. The wide bandwidth is due to the high degree of association of hydrogen with other hydroxyl groups by creating strong bonds between the OH^- and Si ($\equiv Si - OH$) groups. The second significant peak, between 1150 cm^{-1} and 1250 cm^{-1}, can be associated with the band specific to the rhythmic motion along the axis of the covalent bond, which is known as elongation vibration, of the asymmetric group of Si-O -And. The vibration band between 800 cm^{-1} and 700 cm^{-1} is specific to the asymmetric groups of Si-O-Al in the compounds of the analyzed material, and the band between 700 cm^{-1} and 600 cm^{-1} is assigned to the ring tetrahedra of Si-O .

The types of samples obtained, presented in Table 3.6, belong to the broad category of geopolymers for the following reasons:

- is formed as a result of a chemical reaction between a material with a high content of aluminum and silicon oxides and an alkaline solution (based on sodium hydroxide and sodium silicate);
- has a structure similar to zeolitic materials (confirmed by microstructural analysis, NMR and DTA);
- has mechanical characteristics comparable to those of Portland cement-based materials (confirmed by Vicat tests, compression tests and bending tests);
- has a semi-crystalline structure (the diffraction pattern shows areas specific to amorphous materials, but also peaks specific to crystalline phases);
- contain silicon, oxygen and aluminum compounds (sialates) formed as a result of the chemical reaction between the raw material and the alkaline activator (confirmed by FTIR spectra).

Therefore, the four types of geopolymers obtained are semi-crystalline materials, rich in aluminum and silicon oxides, with porous structure and mechanical characteristics comparable to those of Portland cement-based materials in classes B150 and B250, respectively.

References

1.   Davidovits, J., Geopolymer, Green Chemistry and Sustainable Development Solutions ... - Google Books, available online, (accessed on Feb 17, 2020).

2.   Davidovits, J. *30 Years of Successes and Failures in Geopolymer Applications. Market Trends and Potential Breakthroughs.*

3.   Ma, Y.; Hu, J.; Ye, G. The pore structure and permeability of alkali activated fly ash. *Fuel* 2013, *104*, 771–780. https://doi.org/10.1016/j.fuel.2012.05.034

4.   Nergis, D.D.B.; Al Bakri Abdullah, M.M.; Sandu, A.V.; Vizureanu, P. XRD and TG-DTA study of new alkali activated materials based on fly ash with sand and glass powder. *Materials (Basel).* 2020, *13*. https://doi.org/10.3390/ma13020343

5.   Farhana, Z.F.; Kamarudin, H.; Rahmat, A.; Bakri, A.M.M. Al A Study on Relationship between Porosity and Compressive Strength for Geopolymer Paste. *Key*

Eng. Mater. 2013, *594–595*, 1112–1116.
https://doi.org/10.4028/www.scientific.net/kem.594-595.1112

6.   Papa, E.; Medri, V.; Amari, S.; Manaud, J.; Benito, P.; Vaccari, A.; Landi, E. Zeolite-geopolymer composite materials: Production and characterization. *J. Clean. Prod.* 2018, *171*, 76–84. https://doi.org/10.1016/j.jclepro.2017.09.270

7.   Klinowski, J. Nuclear magnetic resonance studies of zeolites. *Prog. Nucl. Magn. Reson. Spectrosc.* 1984, *16*, 237–309.

8.   Ma, L.; Han, L.; Chen, S.; Hu, J.; Chang, L.; Bao, W.; Wang, J. Rapid synthesis of magnetic zeolite materials from fly ash and iron-containing wastes using supercritical water for elemental mercury removal from flue gas. *Fuel Process. Technol.* 2019, 39–48. https://doi.org/10.1016/j.fuproc.2019.02.021

9.   Zeolite synthesis from industrial wastes - ScienceDirect Available online: https://www.sciencedirect.com/science/article/pii/S1387181119303907 (accessed on Nov 21, 2019).

10.   Osacký, M.; Pálková, H.; Hudec, P.; Czímerová, A.; Galusková, D.; Vítková, M. Effect of alkaline synthesis conditions on mineralogy, chemistry and surface properties of phillipsite, P and X zeolitic materials prepared from fine powdered perlite by-product. *Microporous Mesoporous Mater.* 2019. https://doi.org/10.1016/j.micromeso.2019.109852

11.   Dyer, A. Zeolites. In *Encyclopedia of Materials: Science and Technology*; Elsevier, 2001; pp. 9859–9863.

12.   Ismail, A.A.; Mohamed, R.M.; Ibrahim, I.A.; Kini, G.; Koopman, B. Synthesis, optimization and characterization of zeolite A and its ion-exchange properties. *Colloids Surfaces A Physicochem. Eng. Asp.* 2010, *366*, 80–87. https://doi.org/10.1016/j.colsurfa.2010.05.023

13.   Andoni, A.; Delilaj, E.; Ylli, F.; Taraj, K.; Korpa, A.; Xhaxhiu, K.; Çomo, A. FTIR spectroscopic investigation of alkali-activated fly ash: Atest study. *Zast. Mater.* 2018, *59*, 539–542. https://doi.org/10.5937/zasmat1804539a

14.   Khan, S.A.; Uddin, I.; Moeez, S.; Ahmad, A. Fungus-Mediated Preferential Bioleaching of Waste Material Such as Fly - Ash as a Means of Producing Extracellular, Protein Capped, Fluorescent and Water Soluble Silica Nanoparticles. *PLoS One* 2014, *9*, e107597. https://doi.org/10.1371/journal.pone.0107597

15.   Williams, R.P.; Van Riessen, A. Determination of the reactive component of fly ashes for geopolymer production using XRF and XRD. *Fuel* 2010, *89*, 3683–3692.

https://doi.org/10.1016/j.fuel.2010.07.031

16.   Criado, M.; Fernández-Jiménez, A.; Palomo, A. Alkali activation of fly ash. Part III: Effect of curing conditions on reaction and its graphical description. *Fuel* 2010, *89*, 3185–3192. https://doi.org/10.1016/j.fuel.2010.03.051

17.   Torres-Carrasco, M.; Palomo, J.G.; Puertas, F. Sodium silicate solutions from dissolution of glasswastes. Statistical analysis. *Mater. Construcción* 2014, *64*, e014. https://doi.org/10.3989/mc.2014.05213

18.   Criado, M.; Aperador, W.; Sobrados, I. Microstructural and mechanical properties of alkali activated Colombian raw materials. *Materials (Basel).* 2016, *9*. https://doi.org/10.3390/ma9030158

19.   Khan, M.I.; Azizli, K.A.; Sufian, S.; Man, Z.; Khan, A.S.; Ullah, H.; Siyal, A.A. A Short Review of Infra-Red Spectroscopic Studies of Geopolymers. *Adv. Mater. Res.* 2016, *1133*, 231–235. https://doi.org/10.4028/www.scientific.net/amr.1133.231

20.   Ravikumar, D.; Neithalath, N. Effects of activator characteristics on the reaction product formation in slag binders activated using alkali silicate powder and NaOH. *Cem. Concr. Compos.* 2012, *34*, 809–818. https://doi.org/10.1016/j.cemconcomp.2012.03.006

21.   Wan, Q.; Rao, F.; Song, S.; García, R.E.; Estrella, R.M.; Patiño, C.L.; Zhang, Y. Geopolymerization reaction, microstructure and simulation of metakaolin-based geopolymers at extended Si/Al ratios. *Cem. Concr. Compos.* 2017, *79*, 45–52. https://doi.org/10.1016/j.cemconcomp.2017.01.014

22.   Lemougna, P.N.; Nzeukou, A.; Aziwo, B.; Tchamba, A.B.; Wang, K. tuo; Melo, U.C.; Cui, X. min Effect of slag on the improvement of setting time and compressive strength of low reactive volcanic ash geopolymers synthetized at room temperature. *Mater. Chem. Phys.* 2020, *239*. https://doi.org/10.1016/j.matchemphys.2019.122077

23.   Madavarapu, S.B.; Neithalath, N.; Rajan, S.; Marzke, R. *FTIR Analysis of Alkali Activated Slag and Fly Ash Using Deconvolution Techniques*; 2014;

24.   Zhang, S.; Keulen, A.; Arbi, K.; Ye, G. Waste glass as partial mineral precursor in alkali-activated slag/fly ash system. *Cem. Concr. Res.* 2017, *102*, 29–40. https://doi.org/10.1016/j.cemconres.2017.08.012

25.   Anbalagan, G.; Prabakaran, A.R.; Gunasekaran, S. Spectroscopic characterization of indian standard sand. *J. Appl. Spectrosc.* 2010, *77*, 86–94. https://doi.org/10.1007/s10812-010-9297-5

6. Physical-Mechanical Analysis of Geopolymers

6.1. Setting time

The setting time of the geopolymer paste was evaluated by Vicat test according to ASTM C 191-04a. The test was conducted at room temperature (22 ±2 °C) with a Vicat apparatus (Figure 6.1.a).

The Vicat method consists in testing the penetration resistance, with a flat tip needle with a diameter of 1 mm, of a geopolymer sample that was poured into a frustoconical mould specific to the device used (Figure 1. b). After placing the mould on a glass plate, it is filled with freshly obtained geopolymer paste (immediately after mixing the components) and covered with a glass panel that prevents water evaporation. Every 30 minutes, the penetration resistance shall be checked by removing the glass panel and releasing the needle using the release screw. The needle is pushed into the sample surface under the action of its weight, the weight of the plunger and that of the head (300 g). Depending on the depth of penetration, it is determined whether the two points (initial or final) specific to the setting time have been reached. After checking the depth on the scale, the needle is removed from the paste by lifting the rod, further, the sample is covered with the glass panel for the next 30 minutes.

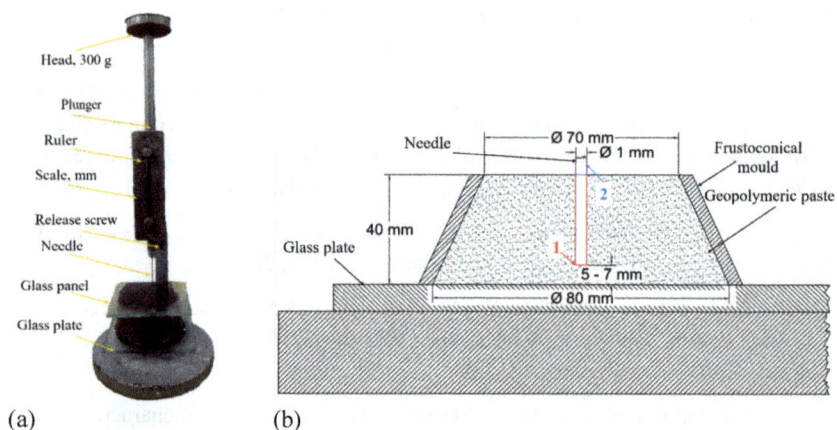

Figure 6.1. Vicat apparatus for setting time measurement: (a) components; (b) schematic representation of measurement principle.

The interpretation of the setting time for these materials is similar to the equilibrium diagram specific to alloy systems, where the cooling liquid line provides information about the temperature at which the first crystallization germs appear, and the solidus line highlights the temperature at which the transition to solid-state occurs for the entire volume of material [1]. In the case of geopolymers, the first point (position 1 of the needle tip) refers to the initial setting time, which is specific to the loss of plasticity of the mixture and the passage of a certain period of time after mixing the components. The second point (position 2 of the needle tip) refers to the final setting time and this is characterized by the total time elapsed from the mixing of the components to the actual hardening of the sample (Figure 6.2). In other words, the initial setting time is determined by the time elapsed from the pouring in the mould of the geopolymeric paste until the needle penetrates to a depth of only $(5 \div 7)$ mm in the geopolymer paste (Figure 1.b). The final setting time is specific to the moment when the needle no longer penetrates the surface of the sample.

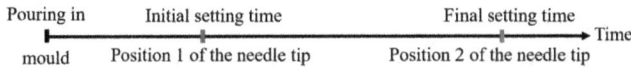

Figure 6.2. Time distribution of the points specific to setting time.

The coal ash-based geopolymers analysed in this study shows an initial setting between 4.78 and 6.88 hours and a final setting time between $22.95 \div 25.00$ hours at ambient temperature (Table 6.1).

Table 6.1. Setting time of coal ash based geopolymers.

Sample	100FA		70FA_30PG		30FA_70S		15FA_15PG_70S	
	a	b	a	b	a	b	a	b
Test 1	290	1420	410	1500	280	1380	350	1410
Test 2	300	1420	410	1490	290	1370	350	1400
Test 3	290	1430	420	1510	290	1380	370	1430
Average, [min.]	293	1423	413	1500	287	1377	357	1414

a – initial setting time, [min.]; b – final setting time, [min.].

It is essential that the initial setting time to be long enough to cover all the cast-in-place operations. However, according to the obtained results, the initial and final setting time of the samples with glass powder is higher than that of 100FA and 30FA_70S samples. In case of this materials calcium silicate hydrate (C-S-H), calcium aluminate silicate hydrate (C-A-S-H) and sodium aluminate silicate hydrate (N-A-S-H) are formed during geopolymerisation. As reported in previous studies the compounds that include calcium have lower setting time because the dissolution rate of Si^{4+} and Al^{3+} species is lower compared to the dissolution rate of Ca^{2+}. Therefore, the setting time of the samples with glass powder should be lower, but this inversion can be explained through the dissolution of high content of Si^{4+}, as a consequence of the glass particles reaction in the alkaline environment, that results in a high initial and setting time.

Due to the fineness of the glass powder, the Si^{4+} content from their surface is dissolved in high concentration by the alkaline solution. Therefore, the setting time of the samples with glass particles (70FA_30PG and 15FA_15PG_70S) show higher initial and final setting time, even with a higher Ca content.

6.2. Relative pore size distribution

Proton nuclear magnetic resonance (NMR) relaxometry is a valuable technique that can be used to extract information about the pore size distribution of porous materials. The technique relies on the proportionality between the pore size and the relaxation time (transverse or longitudinal) of protons belonging to the liquid molecules confined inside pores [2,3]. Thus, from the relaxation time distribution, it is possible to extract the pore size distribution. Note however that, the proportionality between the pore size and relaxation time is valid only if one neglects the bulk relaxation rate of the confined molecules. Moreover, in the case of porous media with magnetic impurities, it is necessary to reduce diffusion effects on transverse relaxation measurements [4,5]. A valuable approach to reduce diffusion effects on transverse relaxation measurements is to implement the well-known Carr Purcell Meiboom Gill (CPMG) technique [6] in combination with a lowfield NMR instrument.

Transverse relaxation measurements of the proton spins confined inside geopolymers were performed using the CPMG technique [7]. Recording of the CPMG echo trains was performed using a low field NMR instrument, operating at 20 MHz proton resonance frequency (Minispec MQ20, Bruker Optics, Germany). The echo time used in our investigations was 0.1 ms, which allowed neglecting of the diffusion effects on echo train attenuation. The NMR measurements were performed first on samples maintained in fresh air at room temperature conditions (21-23 Celsius degrees; 40-48% air humidity) for 48 days to highlight the residual activator in pores. Then all the samples were immersed in water for 7 days and measured again in order to evaluate water absorption and the relative size distribution of all pores in the structure. The relaxation time distribution was obtained from the CPMG echo trains using a numerical Laplace transform [8]. Provided that one can neglect the bulk contribution to the relaxation rate and the interaction of confined molecules with the surface of the investigated samples is identical, then the relaxation time distribution mimics the pore size distribution.

From a structural point of view, the geopolymers contain unreacted raw material due to several factors, such as improper mixing, too low liquid to solid ratio, etc. This can be observed in the form of surfaces covered with spheres (ash particles) in SEM micrographs (Figure 4). However, there is no exact method to quantitatively evaluate the unreacted ash from a specific sample. Due to their spherical structure, the coal-ash particles which do not react before the setting time ending, influence significantly the porosity of the geopolymers.

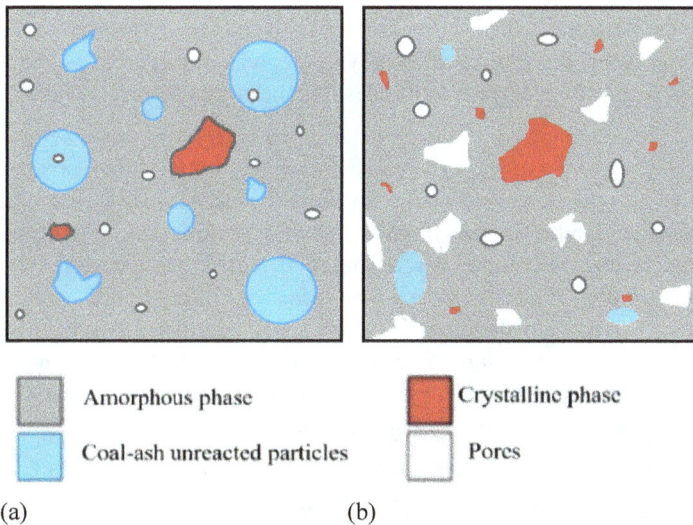

Amorphous phase   Crystalline phase

Coal-ash unreacted particles   Pores

(a)   (b)

Figure 6.3. Schematic representation of geopolymers morphology: (a) after setting time ending; (b) after reacting [9].

Following the formation of the solid structure (Figure 6.3. (a)), the regions inside the samples containing ash particles and activation solution will react in time creating pores of different sizes. Also, the pores formation or their increase in size could be affected by the curing time increase, because the conversion of the amorphous phase into the crystalline phase occurs during the curing stage (Figure 6.3. (b)).

The CPMG series collected from the NMR measurements were used to obtain the relaxation time distributions (T_2) as presented in the following figures. The resulting peaks were analyzed by a comparison between different types of samples. The graphs specific to the samples maintained in the room conditions (21-23 Celsius degrees; 40-48% air humidity) for 48 days (Figure 6.4.a), Figure 6.5.a), Figure 6.6.a) and Figure 6.7.a)) reveal two peaks. The first peak can be assigned to the liquid protons (residual activator solution) resulting from the geopolymerization process in the partially filled gel pores. The second peak can be attributed to the water absorbed inside the capillary pores from the atmosphere or can be an artifact of the numerical Laplace inversion. It is known that numerical inverse Laplace is ill-conditioned and may lead to spurious peaks when applied to noisy data. Comparing the data on the samples maintained at room conditions with those of the samples immersed in water for 7 days, one can observe a significant increase of the peak area which is proportional to the number of protons confined inside the pores. It is also observed that different types of pores cannot be graphically separated due to the rapid exchange of water molecules from one type of pore to another.

The graph of the sample 100FA (Figure 6.4.b) dried for 8 hours shows several peaks on the curve (black dots curve), the first peak between ≈0.1 ms and 1 ms corresponds to the liquid in the gel-type pores (<50 nm), the second peak between ≈ 1 ms and 7.5 ms corresponds to the capillary pores (50 - 600 nm), and the third peak > 7.5 ms corresponds to the liquid in the voids (pores larger than 600 nm) or the cracks results from the crystalline phase growth . As the drying time increases, the gel remaining on the surface of the unreacted or partially dissolved coal-ash particles continues to activate, resulting in the ash spheres opening, therefore, pores volume increases. At the same time, the density of the sample decreases as a result of the structure permeability increase due to the remaining water elimination from the small pores.

As can be seen in Figure 6.4.b), the effect of drying time increasing, from 8 hours to 16 hours, on the gel pore size distribution is minimum, but in the case of 24 hours dried samples the amount of liquid detected is much smaller. This phenomenon can be explained by the decrease in the number of gel pores following the reaction between the activator and the unreacted ash particles, i.e. pore growth. Therefore, the increase of the structure dehydration degree as a result of the drying time increase produces a decrease in the total volume occupied by gel-type pores.

When replacing 30% of the coal-ash with glass particles smaller than 10 μm in diameter, a significant decrease in gel pores occurs, mostly because the sample volume is filled with compact particles. As the drying time increases, the number of pores specific to the first peak decreases significantly, but the characteristic curve of the sample maintained for 24 hours shows an additional peak between 0.54 ms and 2.51 ms (Figure 6.5.b). This additional peak can be explained by the gel pores connection following activation of the unreacted ash forming a category of intermediate pores. Also, it can be observed that with the increase of the drying time the curves become more flattened and the transition from one category of pores to another is less visible. This phenomenon can be related to the increase of the structure permeability and also to the pores size distribution.

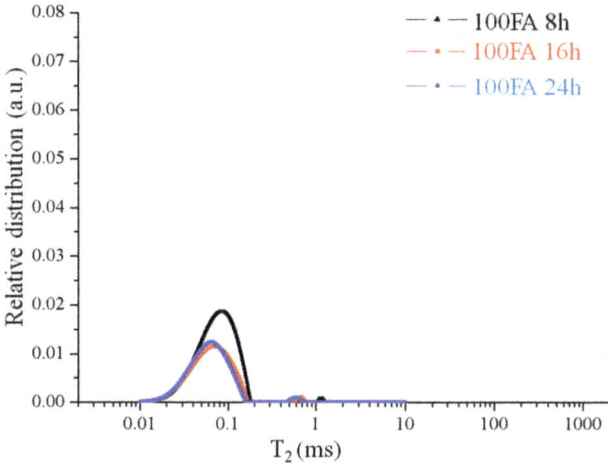

(a)

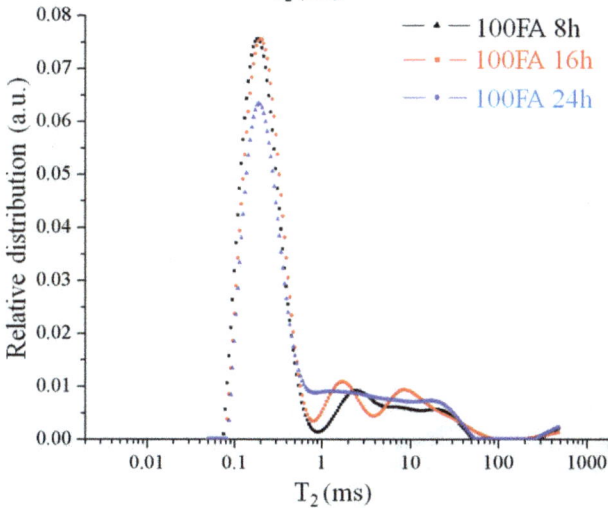

(b)

*Figure 6.4. Relative pore size distribution in sample 100FA dried for 8, 16 or 24 hours:
(a) after 48 days of activation; (b) after 7 days of immersion in water.*

The geopolymers samples with 70% by mass sand of solid components present a higher decrease in the number of gel-type pores (Figure 6.6.b). When the drying time is increased the number of gel-type pores decreases even higher due to the conversion of gel pores in capillary pores and capillary pores in large pores (voids). Therefore, the drying time positively influences the growth of the pores as a result of the reaction between the undissolved ash particles and the activation solution from the gel pores.

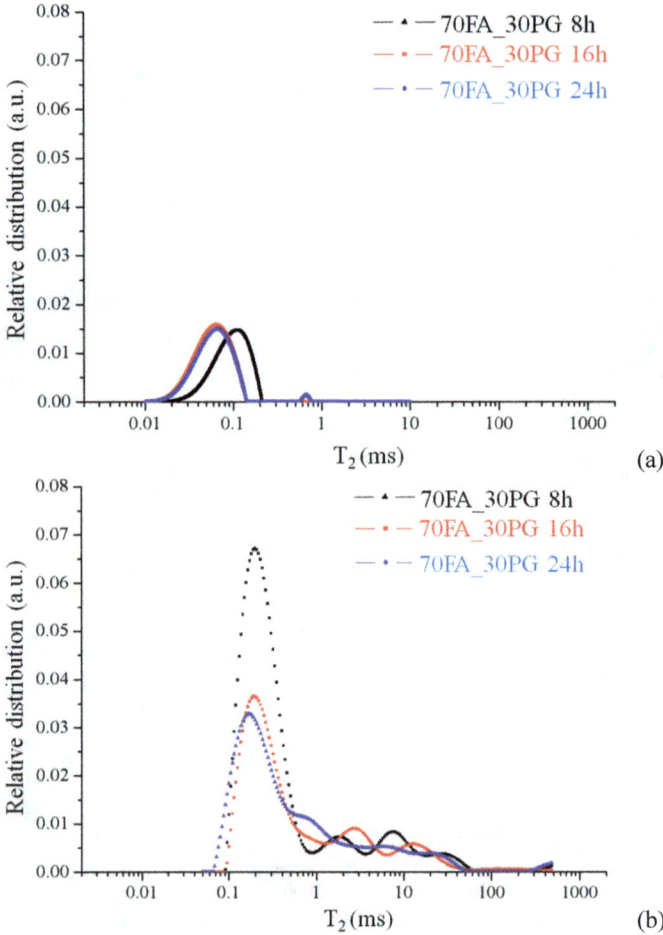

(a)

(b)

*Figure 6.5. Relative pore size distribution in sample 70FA dried for 8, 16 or 24 hours:
(a) after 48 days of activation; (b) after 7 days of immersion in water.*

However, the lowest pores size distribution was obtained for the samples with 15% ash, 15% glass powder and 70% sand due to the percentage of compact particles in the analyzed sample volume. Therefore, the coal-ash percentage from the solid component is directly proportional to the matrix volume, its replacement with compact particles results in the decrease of gel pores number (Figure 6.7.b).

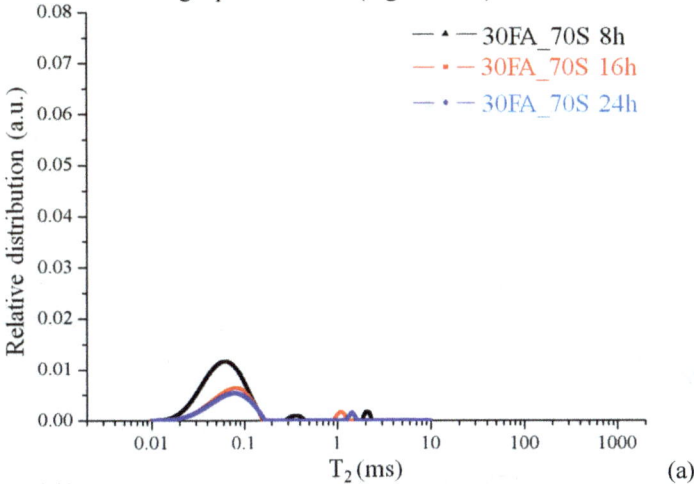

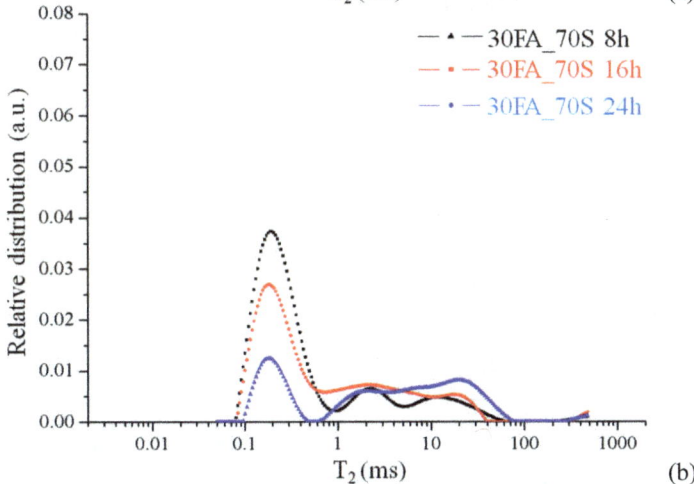

Figure 6.6. Relative pore size distribution in sample 30FA dried for 8, 16 or 24 hours: (a) after 48 days of activation; (b) after 7 days of immersion in water.

Also, the type of aggregate influences the geopolymers microstructure, by introducing the glass powder into the composition, a decrease of gel pores occurs, but the relative distribution of capillary and large pores is approximately the same. However, when sand is introduced, the gel pores relative distribution decreases, but that of capillary and large pores increases.

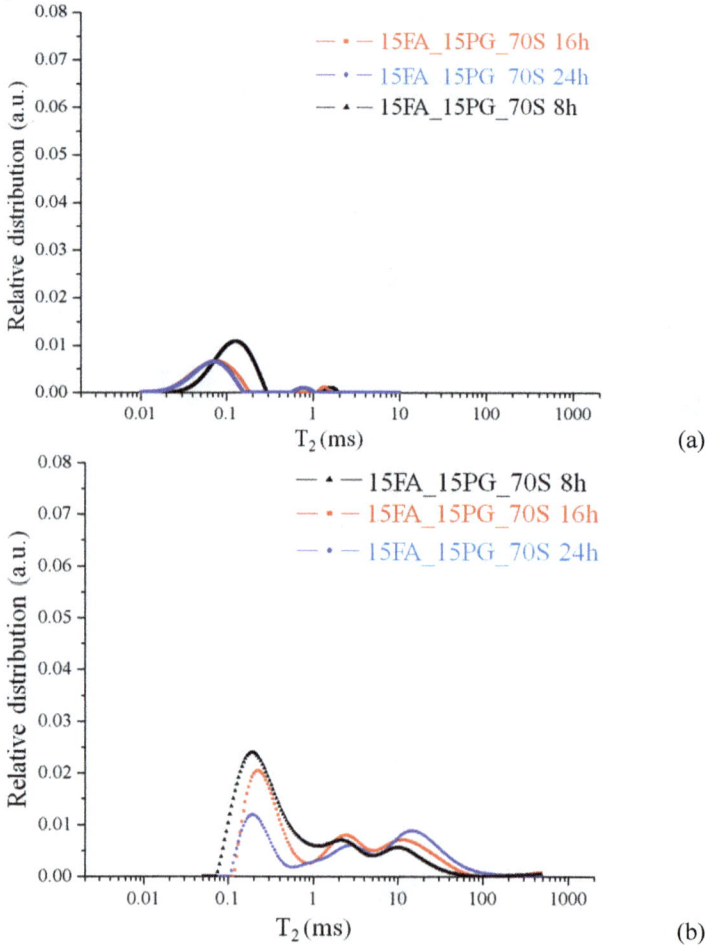

Figure 6.7. Relative pore size distribution in sample 15FA dried for 8, 16 or 24 hours: (a) after 48 days of activation; (b) after 7 days of immersion in water.

When the pore size relative distributions are compared depending on the sample composition for 8 hours (Figure 6.8.a), 16 hours (Figure 6.8.b) and 24 hours (Figure 6.8.c) drying time, all the curves present three peaks which area is decreased by the increase of the percentage of the reinforcing particles. The first peak with the smallest area decreases as a result of the coal-ash percentage decreasing. However, the influence of the particles on the relative distribution of capillary and large pores is relatively low. The area values and peaks position (X_1 and X_2) on the T_2 axis are presented in Table 6.2.

Table 6.2. Peaks areas and positions on the T_2 axis.

Sample	Drying time	Peak 1			Peak 2			Peak 3		
		X_1 (ms)	X_2 (ms)	Area (a.u.)	X_1 (ms)	X_2 (ms)	Area (a.u.)	X_1 (ms)	X_2 (ms)	Area (a.u.)
15FA	8 h	0.07	1.07	0.01	1.07	5.17	0.02	5.17	86.81	0.11
30FA	8 h	0.07	0.91	0.01	0.91	5.24	0.01	5.24	61.16	0.12
70FA	8 h	0.09	0.89	0.01	0.89	3.71	0.01	3.71	63.17	0.18
100FA	8 h	0.07	0.91	0.02	0.91	5.63	0.03	5.62	50.40	0.17
15FA	16 h	0.11	0.88	0.01	0.88	5.08	0.02	5.08	76.65	0.20
30FA	16 h	0.08	0.69	0.01	0.69	9.91	0.05	9.91	40.40	0.10
70FA	16 h	0.09	1.25	0.01	1.25	6.42	0.03	6.42	60.60	0.12
100FA	16 h	0.07	0.80	0.02	0.80	3.83	0.02	3.83	91.74	0.24
15FA	24 h	0.10	0.53	0.01	0.53	4.94	0.02	4.94	150.2	0.35
30FA	24 h	0.09	0.56	0.01	0.56	5.57	0.02	5.57	76.65	0.30
70FA	24h	0.06	1.65	0.01	1.65	12.65	0.05	12.65	53.02	0.09
100FA	24h	0.08	0.82	0.01	0.82	10.37	0.07	40.37	55.78	0.18

The relative pore size distribution of the coal-ash geopolymers confirms the presence of three types of pores in the geopolymers structure. These reveal a high size distribution, ranging from nanometers (gel or capillary pores) to millimeters (large pores or voids). The relative distribution between these three types of pores is influenced by the drying time and also by the percentage of reinforcing particles. Therefore, by increasing the drying time, the gel remaining on the surface of the unreacted or partially dissolved coal-ash particles continues to activate, resulting in the ash spheres opening and pores volume increase. Moreover, by increasing the percentage of the reinforcing particles the number of gel-type pores decreases proportionally. Therefore, the lowest gel-type pores size distribution was obtained for the samples with 15% ash, 15% glass powder and 70% sand due to the percentage of compact particles in the analyzed sample volume. In other words, the coal-ash percentage from the solid component is directly proportional to the matrix volume, its replacement with compact particles results in the decrease of gel pores number.

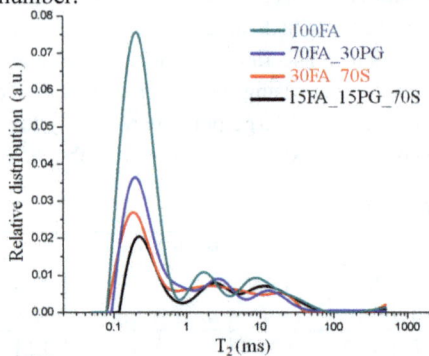

(a)

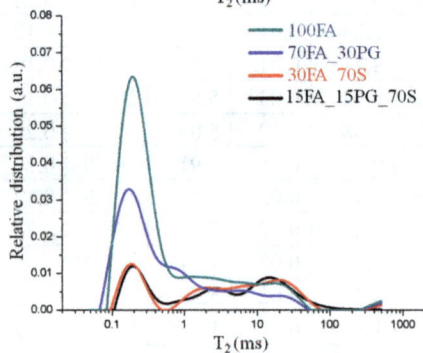

(b)

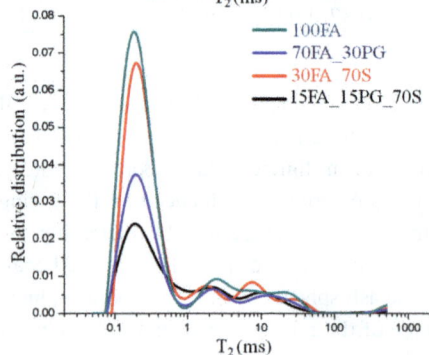

(c)

Figure 6.8. Relative pore size distribution of obtained geopolymers by drying time:
(a) 8 hours; (b) 16 hours; (c) 24 hours.

Also, the type of aggregate influences the geopolymers microstructure, by introducing the glass powder into the composition, a decrease of gel pores occurs, but the relative distribution of capillary and large pores is approximately the same.

According to this study, the obtained geopolymers contain three types of pores: gel (<50 nm), capillary (50 - 600 nm) and large pores (pores greater than 600 nm) formed by the arrangement of the OH⁻ and Si groups (Si-OH), Si-O-Si groups, Si-O-Al groups and Si-O rings.

6.3. Compressive strength

In the case of oxide materials, under the action of an external force, cracks appear that do not propagate in a straight line, but follow a sinusoidal direction around the reinforcing particles or around the constituents from the structure. In the analysed geopolymers, the direction of cracks propagation is strongly influenced by the presence of aggregates (sand or glass particles) or pores, which can block or deviate their formation [10,11].

The compressive strength evaluation of the obtained geopolymers was performed according to the Standard C109/C109M-07 requirements, on a batch of at least 5 specimens with dimensions of 50x50x50 mm³. According to the tests, by replacing 70% by weight of power plant ash with aggregates, the compressive strength increases from 8 MPa to 12 MPa for 7-day samples, from 12 MPa to 20 MPa for 28-day samples and from 15 MPa to 27 MPa for 90-day samples. The compressive strength values presented (Figure 6.9) have been calculated as the arithmetic mean of the values obtained on the entire bach.

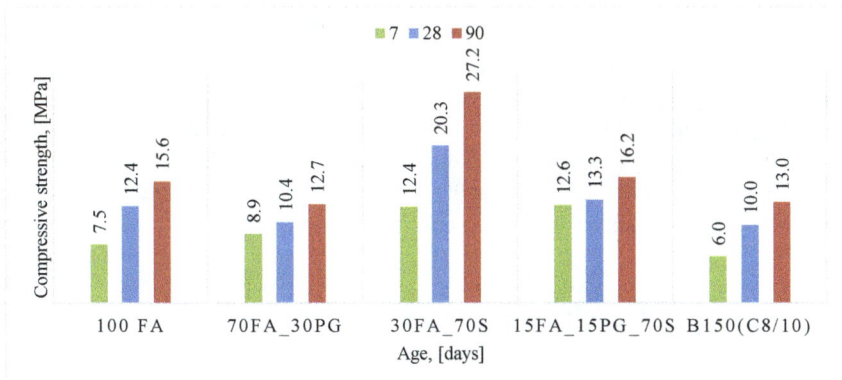

Figure 6.9. Compressive strength resistance at 7, 28 and 90 days.

In terms of compressive strength, according to the standard NE 012-99, the 100FA, 70FA_30PG and 15FA_15PG_70S geopolymers belong to class B150 or C8/10 of Ordinary Portland Cement based concrete. This class is mainly used for pouring leveling layers to foundations. Instead, the 30FA_70S geopolymer belongs to class B250 or C16 / 20 and can be used for the execution of resistance structures (pillars, beams, belts or plates), foundations or as a support layer for reinforced concrete. The geopolymer belonging to B250 class is a suitable material, in terms of compressive strength, to replace OPC based concrete in ground floor constructions plus a floor. Moreover, due to the size of the particles (<16 mm in diameter) from the composition, the obtained geopolymers can be cast by concrete pumps.

6.4. Flexural strength

The flexural strength was tested, according to Standard SR EN 196-1 requirements, by applying a force at the center point on the upper surface of a sample, when the lower surface is placed on two supports. The flexural strength test was performed on 40x40x160 mm^3 specimens according to the three-point method [12]. Figure 6.10 shows the differences in bending strength between the coal ash based geopolymers and the samples with reinforcing particles in composition. The difference between the values (Figure 6.10) obtained for the samples 100FA and 70FA_30PG tested at the age of 7 and 28 days is less than 0.6 MPa ($\approx$11%), but at 90 days the differences increases up to 2.1 MPa ($\approx$25%). At the age of 90 days, the difference between the 100FA and 30FA_70S sample is of approximately 0.6 MPa (6%). Thus, it can be seen that by introducing the glass powder there is a decrease in bending strength at an later age. The phenomenon is due to the dissolution of glass particles that increase the Si content available in the system, thus changing the Si: Al ratio and producing a more fragile matrix.

The flexural strength of the obtained geopolymers was (27.5 ÷ 42.5)% higher at 7 days, by (26.0 ÷ 74.0)% higher at 28 days, respectively by (9.4 ÷ 45.71)% higher at 90 days than that of concrete of class B150. Also, due to the dehydration of the matrix, the interface between the unreacted area and the matrix (Figure 5.16) increased with the age of the samples, thus, it was found that at 90 days the bending strength of the 100 FA sample is higher than samples containing reinforcing particles.

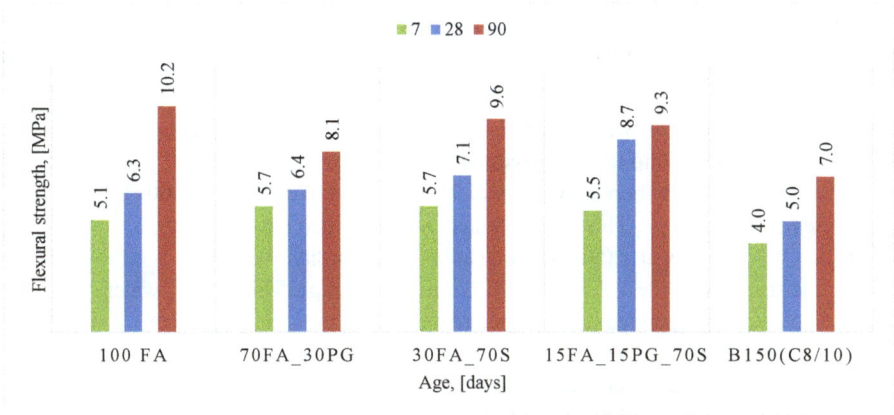

Figura 6.10. Rezistența la încovoiere a geopolimerilor la 7, 28 și 90 de zile.

6.5. Thermal behavior evaluation of geopolymers

The evaluation of the mass evolution of the geopolymer samples by thermogravimetric analysis (TG) simultaneously with the study of the changes of the chemical structure by differential thermal analysis (DTA) was performed with the help of STA PT-1600 equipment. The evaluation of the thermal behavior was performed in the temperature range $(25 \div 800)$ °C, with a heating rate of 10 °C/min on samples with a mass of less than 50 mg in air atmosphere.

Materials analysis by TG-DTA emphasizes their thermal stability and the content/type of volatile compounds through two curves simultaneously plotted according to the temperature. The curve between the temperature of the enclosure and the heat flow (temperature) difference between the reference sample and the analyzed sample, measured in µV, highlights the structural changes. While, the curves between the chamber temperature and the sample mass shows a decrease or increase of the analyzed sample mass, measured in milligrams (mg).

The TG-DTA simultaneous thermal analysis was used to evaluate the thermal stability of geopolymers after replacing high percentages of fly ash with two types of particles. By monitoring the mass change during heating of samples, the fraction of volatile compounds can be determined, so if the DTA curve is plotted at the same time the mass change at specific temperatures can confirm the quantity of a specific compound.

The DTA curves of samples show multiple peaks at 123-130 °C, 185 °C, 232-240 °C, 312-358 °C, 490-497 °C and 572-576 °C, respectively (Table 6.3). These peaks correspond to water molecules remove, which are free or bound with the structural compounds. In totally inorganic materials, such as geopolymers, water can be found in two main forms:

- hygroscopic (free) water which is removed up to 120 °C [13]. This is absorbed in the structure due to the hygroscopicity of geopolymers.
- strong physically bonded water which is removed in the 120 °C-300 °C temperature range. This type of water can be divided into three types:
- crystallization water (anionic and cationic or coordinative) which is removed from the structure in the 120 °C-200 °C temperature range. This type of water molecules are bonded in the structure during the formation of crystals from aqueous solution [14].
- water from hydrogels that can be intercrystalline and network type which interact with the crystallization water. This type of water is removed during heating in the 180 °C-300 °C temperature range [15].
- zeolitic water from cavities and channels which is removed from the structure in the 200 °C – 300 °C temperature range [16].

Table 6.3. Peaks position on DTA curves.

Sample Peak	Sample 100FA (1)	Sample 70FA_30PG (2)	Sample 30FA_70S (3)	Sample 15FA_15PG_70S (4)
A	123 °C	115 °C	130 °C	130 °C
B	185 °C	-	-	-
C	235 °C	232 °C	233 °C	240 °C
D	320 °C	330 °C	328 °C	330 °C
E	497 °C	490 °C	-	495 °C
F	575 °C	576 °C	572 °C	576 °C

When the temperatures exceed 300 °C, the chemically bound water starts being removed. The peaks showed on the DTA curve above this temperature, corresponds to the decomposition of M (metal) and OH groups compounds [17,18]. These compounds exist in the fly-ash based geopolymers structure in different forms, such as:

- Acids: $M\text{-}O^-H^+$ [Si(IV), Ti(IV), Fe(III)];
- Basics: M^+HO^- hydroxide [Na, Ca (II), K, Mg (II)];
- Neutral: M-OH hydroxyl [Al (III), Mn (III)].

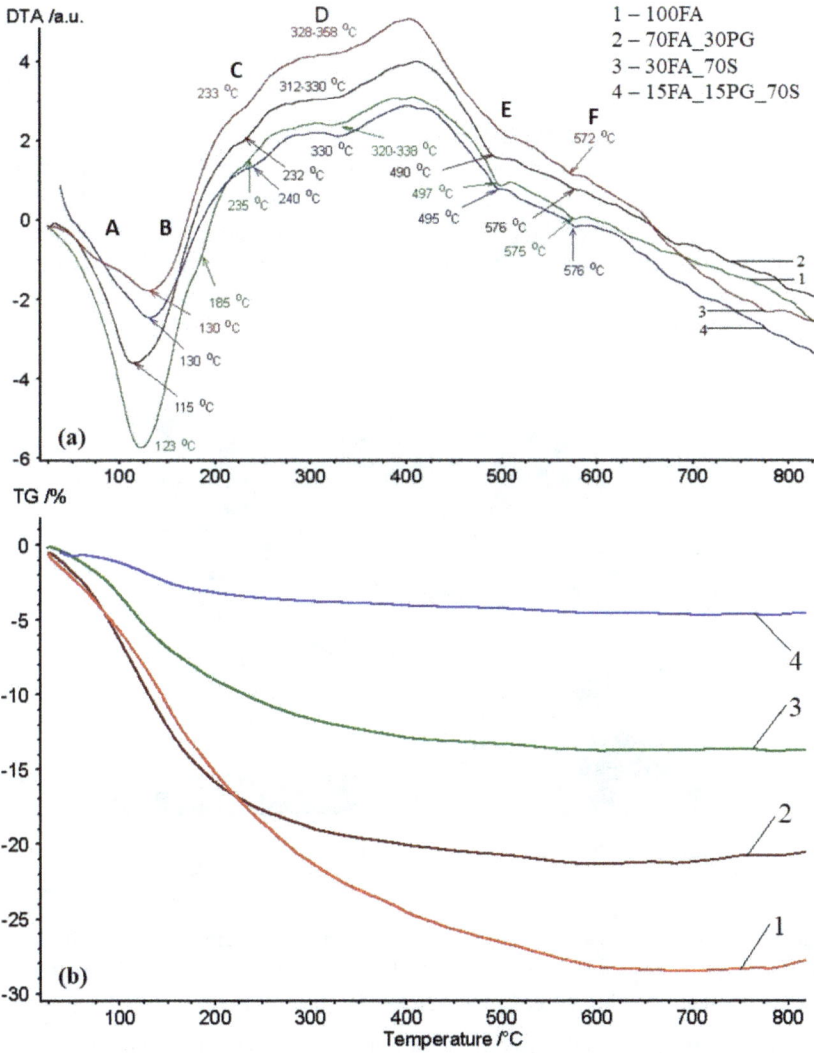

Figure 6.11. TG-DTA curves in the 22 °C – 820 °C temperature range: (a) DTA curves; (b) TG curves.

The DTA curves (Figure 6.11.a) of the analyzed samples show an endothermic peak which minimum is positioned at 123 °C for 100FA sample, 115 °C for 70FA_30PG sample, 130 °C for 30FA_70S and 15FA_15PG_70S, respectively. The peak "A" corresponds to the overlapping of the removing of hygroscopic water evaporation and crystallization water removing [16]. By comparing the peaks broadening it can be seen that by increasing the percentage of compact particles, the amount of water in these forms is lower. Because the used particles are compact bodies (Figure 6.3), the porosity of the sample can be related only with the percentage of fly ash. Therefore, high fly ash content ensures a highly porous structure which will increase the amount of absorbed water. The "B" peaks which are in the temperature range of hydrogel water removing is higher in the case of 100FA sample. This can be related to the hydrogel-forming capability of fly ash during geopolymerisation [19].

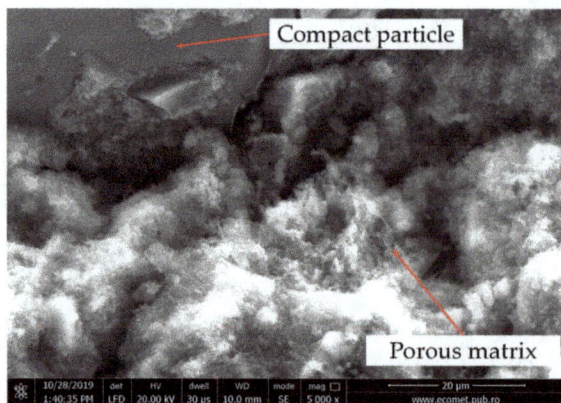

Figure 6.12. SEM micrographs of geopolymers with particles.

Close to 230 °C another peak, "C", appears. During this endothermic reaction the water molecules are removed from the calcium silicate hydrate (C-S-H), C-S-H with Al in its structure (C-A-S-H) and sodium aluminosilicate hydrate (N-A-S-H) channels and pores [20,21].

The "D" peaks correspond to the iron oxides transition from FeO(OH) amorphous phase (Goethite) into the α-Fe_2O_3 (Hematite) crystalline phase (equation 6.1) [22]. The transformation reaction of Fe compounds occurs around 300 °C but could be moved to higher temperatures due to the presence of silica and aluminum [23].

The "E" peaks represents an endothermic reaction with appears in the 490 °C – 497 °C temperature range and corresponds to the calcium hydroxide $Ca(OH)_2$ (Portlandite) (Figure 6.11) decomposition following the reaction with the carbon from the atmosphere resulting in $CaCO_3$ and H_2 (equation 6.2) [1,24].

Also, up to 570 °C, the "F" peaks which appear on the DTA curve corresponds to the α-quartz to β-quartz conversion and the reaction between the unreacted particles and the activator caught in gel pores [25]. However, in the same temperature range aluminium hydroxide, $Al(OH)_3$, decomposition occurs (equation 6.3) [2,26].

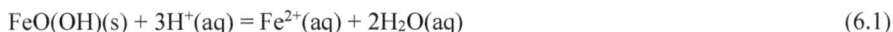

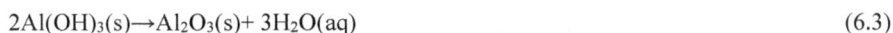

$$FeO(OH)(s) + 3H^+(aq) = Fe^{2+}(aq) + 2H_2O(aq) \tag{6.1}$$

$$Ca(OH)_2(s) + CO(g) \rightarrow CaCO_3(s) + H_2(aq) \tag{6.2}$$

$$2Al(OH)_3(s) \rightarrow Al_2O_3(s) + 3H_2O(aq) \tag{6.3}$$

Anyway, in the same temperature range, the water resulting from the silicon or aluminum hydroxide groups condensation could take place. According to [27] the chemical reaction consists of (equation 6.4):

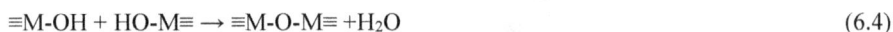

$$\equiv M\text{-}OH + HO\text{-}M \equiv \rightarrow \equiv M\text{-}O\text{-}M \equiv + H_2O \tag{6.4}$$

As can be seen in Figure 6.13.a), the obtained geopolymers present large pores distributed on the entire analyzed surface. After introducing the aggregates, the large pores, especially, decrease in number (Figure 6.13.b).

The samples mass loss (Figure 6.11.b) up to hygroscopic water evaporation is close to 10% for 100FA sample, 8% for 70FA_30PG, 5% for 30FA_70S, and 2% for 15FA_15PG_70S respectively. The water absorbance capacity of samples is related to the calcium oxides and silica gel concentration. Up to the temperature of hydrogels water removing, the samples mass decreases close to 16% in the case of 100FA and 70FA_30PG samples, while the samples with sand show lower than 10% mass reduction. Yet, up to 250 °C the mass decreases close to 18% in the case of 100FA and 70FA_30PG samples, 12 % in the case of 30FA_70S and only 3% in the case of 15FA_15PG_70S.

Even if the percentage of mass loss up to this temperature is relatively high, considering that these types of water molecules are free or physically bonded, their influence on the mechanical properties is insignificant. However, if these materials would be subjected to freeze-thaw cycles, cracks formation may occur due to the water (ice) from the expansion of the pores which will reduce the mechanical resistance of the geopolymers [28].

Figure 6.13. Optical micrographs of: (a) sample 100FA; (b) sample 30FA_70S.

In the 360 °C – 700 °C temperature range, the mass loss is due to the removing of chemical bound water molecules. Therefore, between close to 460 °C and 515 °C which corresponds to 1% mass reduction of 100FA sample and close to 0.2% mass reduction of 15FA_15PG_70S sample is related to the CaOH decomposition. Also, in the $Al(OH)_3$ and $FeO(OH)$ decomposition temperature range, the samples mass loss are lower than 1%.

Furthermore, above this temperature range, the DTA curves still show small peaks, these endothermic or exothermic reactions correspond to the decomposition of $CaCO_3$ at close to 750 °C [29] (equation 6.5), $Ti(OH)_4$ close to 790 °C [30] or $Mg(OH)_2$ close to 670 °C [31]. However, these compounds exist only at tracks level, therefore, the effects on sample characteristics are low. Over 700 °C a mass gain can be observed, this phenomenon appears due to the oxidation of oxygen-poor iron species or pure iron [32].

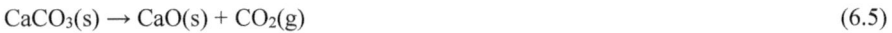

$$CaCO_3(s) \rightarrow CaO(s) + CO_2(g) \tag{6.5}$$

The 100FA sample shows four peaks with the largest area, therefore, the compounds that decompose in the analyzed temperature range come from the power plant fly ash particles.

Microstructural analysis of the samples exposed to high temperature exposed show a conversion of the geopolymers matrix from a grain formed structure in a dense glass-like structure i.e. a sintering process.

Sintering is the process of agglomeration, densification and recrystallization, by thermal activation of the agglomerated mass, in the absence or presence of the molten phase. Sintering is a complex process based on the phenomenon of mass transfer under the action of temperature and pressure (in some cases). During sintering the diffusion and mass transfer process are controlled by the thermal energy. In geopolymers, when the agglomerate of powders and grains begins to sinter, bridges are formed between the adjacent granules, which gradually join until the intergranular spaces are eliminated as advanced as possible [33].

In the absence of the molten phase, the sintering of the ceramics occurs by moving the particles towards each other, reducing their surface and eliminating the pores; all these phenomena lead to an increase in the density of the material. In the first stage of sintering, densification occurs by moving the crystals, by translational and rotational movements, due to high temperature and possibly pressure. Diffusion processes take place in this stage in the volume of the material, and the shape of the particles changes from circular to polygonal - concave, with the reduction of the specific surface of the granules. In this stage an increase of the density in the mass of the material can be observed.

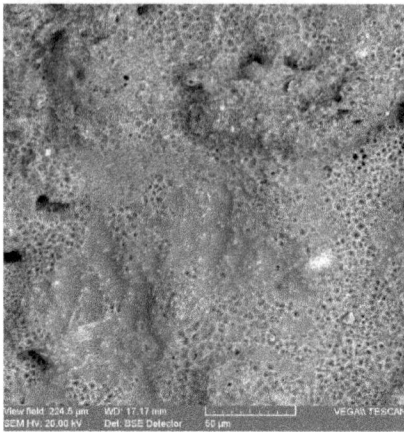
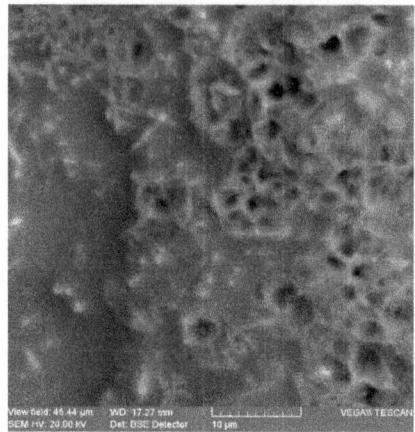

BSE_1KX   *BSE_5Kx*

Figure 6.14. Morphology of 100FA sample after exposure to 1000 °C.

Table 6.4. Chemical composition and elemental distribution in 100FA sample after exposure to 1000 °C.

Element	Si	Na	Fe	Al	Ca	K	Ti	Mg	O
[%], wt.	22.38	10.65	3.88	8.92	2.75	2.30	1.16	0.87	balance
St. error, [%]	1.06	0.80	0.21	0.51	0.15	0.14	0.10	0.11	-

In the second stage of sintering, all the grains are in contact with each other, so their movement is stopped. The contraction of the material can be done only by diffusion through the intergranular spaces or through the existing pore channels at the granular interface. At the end of the second stage, the maximum density of the material will be reached, and the existing pores become very small and closed (isolated). In the third stage of sintering the existing pore shrinks even more due to the growth of crystalline grains, any gases contained in the pores have very high pressure, are insoluble in the solid phase and do not allow the pore volume to shrink.

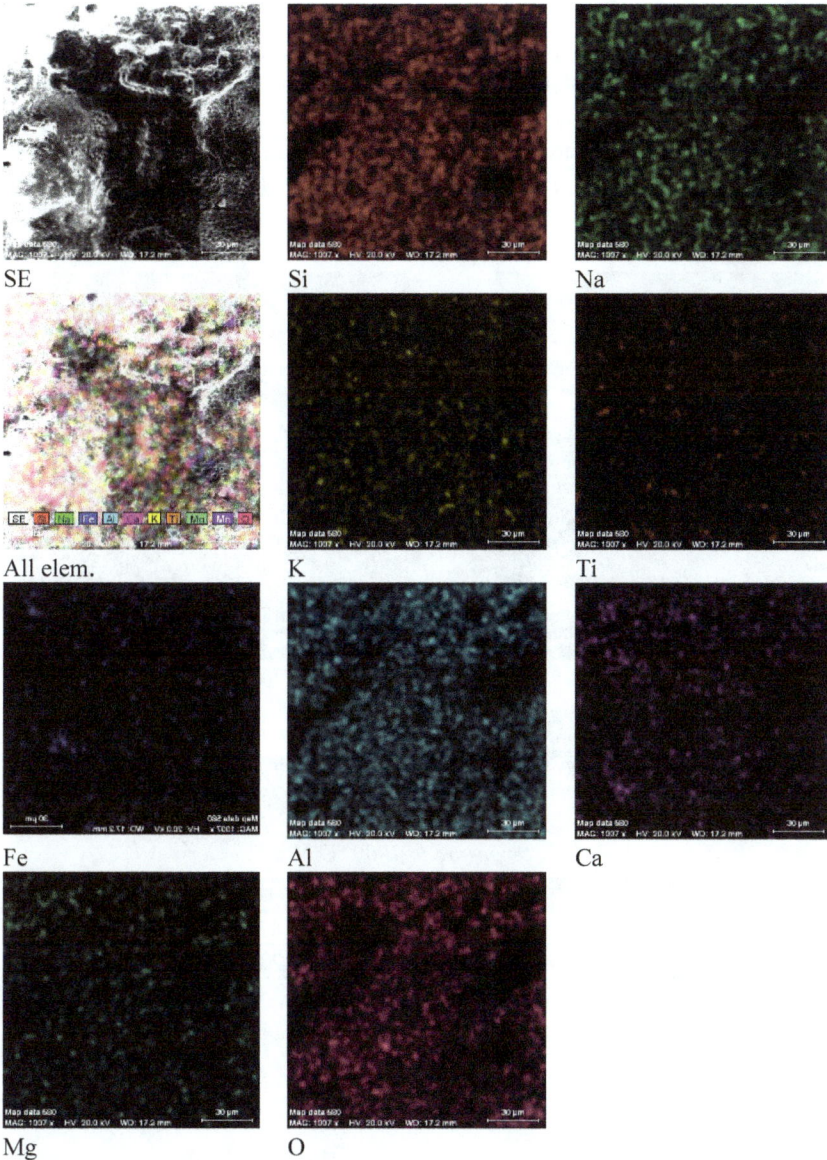

Figure 6.15. Elemental mapping of 70FA_30PG sample.

In the sintering process, the selective recrystallization of the ceramic particles takes place, in parallel with the densification process. Recrystallization consists in the tendency to restore the crystal lattice of the natural ceramic mass, which has been damaged by manufacturing processes (grinding powders, pressing, firing, etc.). The driving force of recrystallization is the difference between the free energy of the granulated and recrystallized material. The small material, so the ceramic powder, has a higher surface free energy than a agglomerated material.

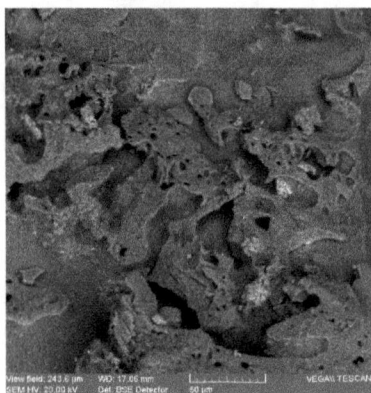 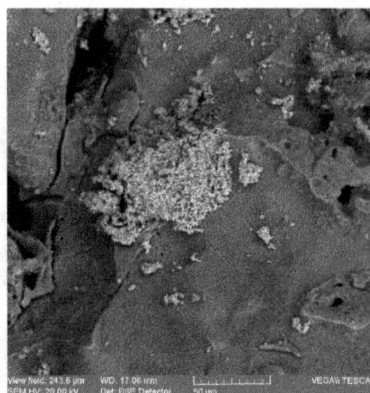

BSE_1KX   *BSE_5Kx*

Figure 6.16. Morphology of 70FA_30PG sample after exposure to 1000 °C.

Table 6.5. Chemical composition and elemental distribution in 70FA_30PG sample after exposure to 1000 °C.

Element	Si	Na	Fe	Al	Ca	K	Ti	Mg	O
[%], wt.	24.19	9.19	4.49	5.07	9.80	1.87	1.57	2.19	balance
St. error, [%]	1.44	1.02	0.44	0.49	0.59	0.23	0.23	0.25	-

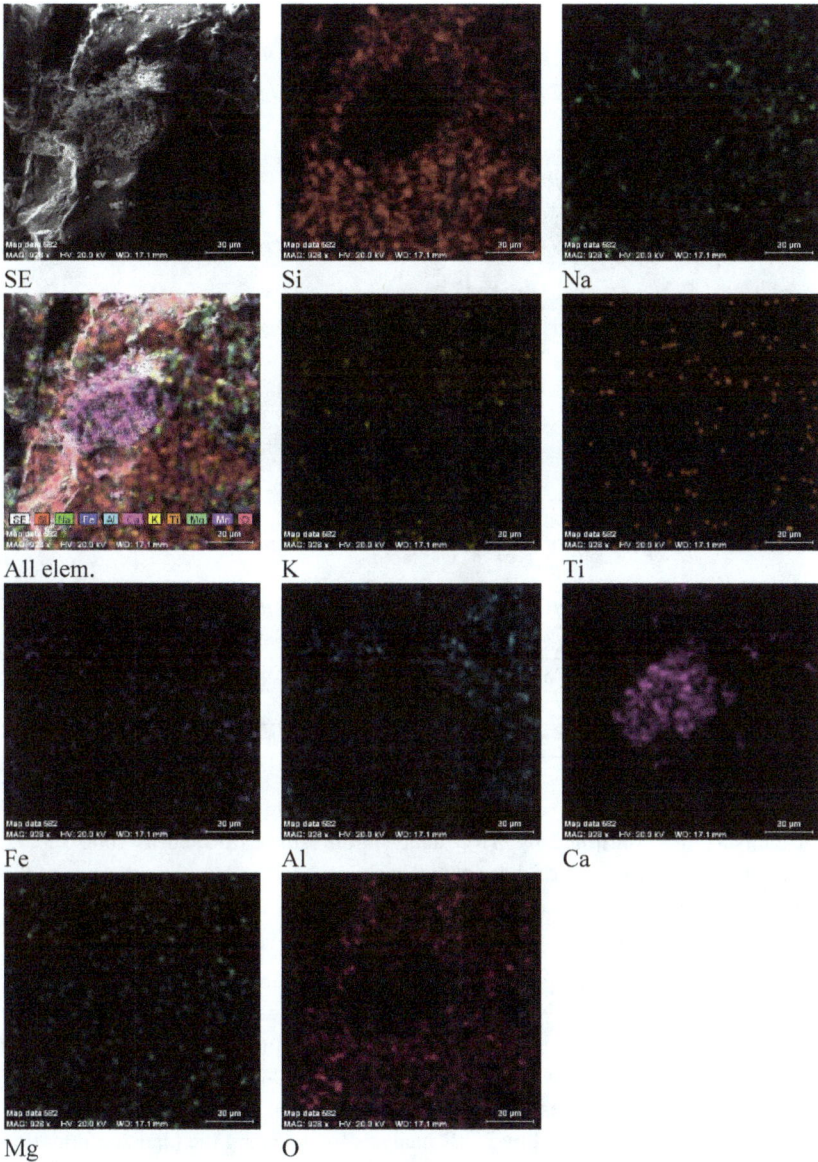

Figure 6.17. Elemental mapping of 70FA_30PG sample.

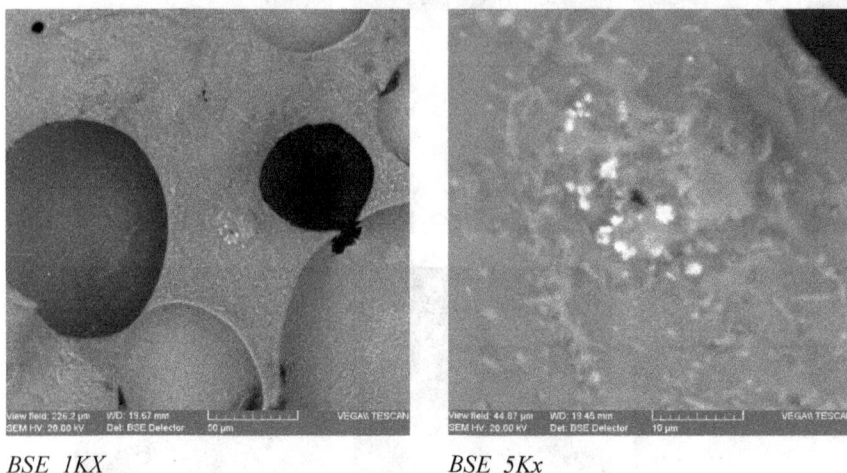

BSE_1KX   BSE_5Kx

Figure 6.18. Morphology of 30FA_70S sample after exposure to 1000 °C.

Table 6.6. Chemical composition and elemental distribution in 30FA_70S sample after exposure to 1000 °C.

Element	Si	Na	Fe	Al	Ca	K	Ti	Mg	O
[%], wt.	30.42	6.61	2.39	6.35	2.35	1.62	0.79	0.89	balance
St. error, [%]	1.71	0.68	0.24	0.49	0.19	0.16	0.12	0.15	-

The decomposition of calcium hydroxide and goethite during heating was identified by TG-DTA analysis and confirmed by XRD analysis on samples heated to 1000 ±5 °C, respectively. The diffractograms of these samples show a decrease in the intensity of the maximum points specific to calcium hydroxide and goethite, as well as a sharp increase in hematite and calcium carbonate.

In the analyzed temperature range (25 ÷ 340) °C, the geopolymer samples show an inflection point, specific to the evaporation of free or physically bound water from the composition. Evaporated water highlights the hygroscopic nature of geopolymers and can be correlated with the amount of water detected by NMR in C-S-H, C-A-S-H and N-A-S-H pores in the structure of samples kept in the air for 48 days. The hygroscopic character of geopolymers was also confirmed by FTIR analysis by detecting the plane deformation vibration of the angle formed between hydrogen and oxygen atoms in the δ-H-O-H bond, specific to adsorbed water molecules.

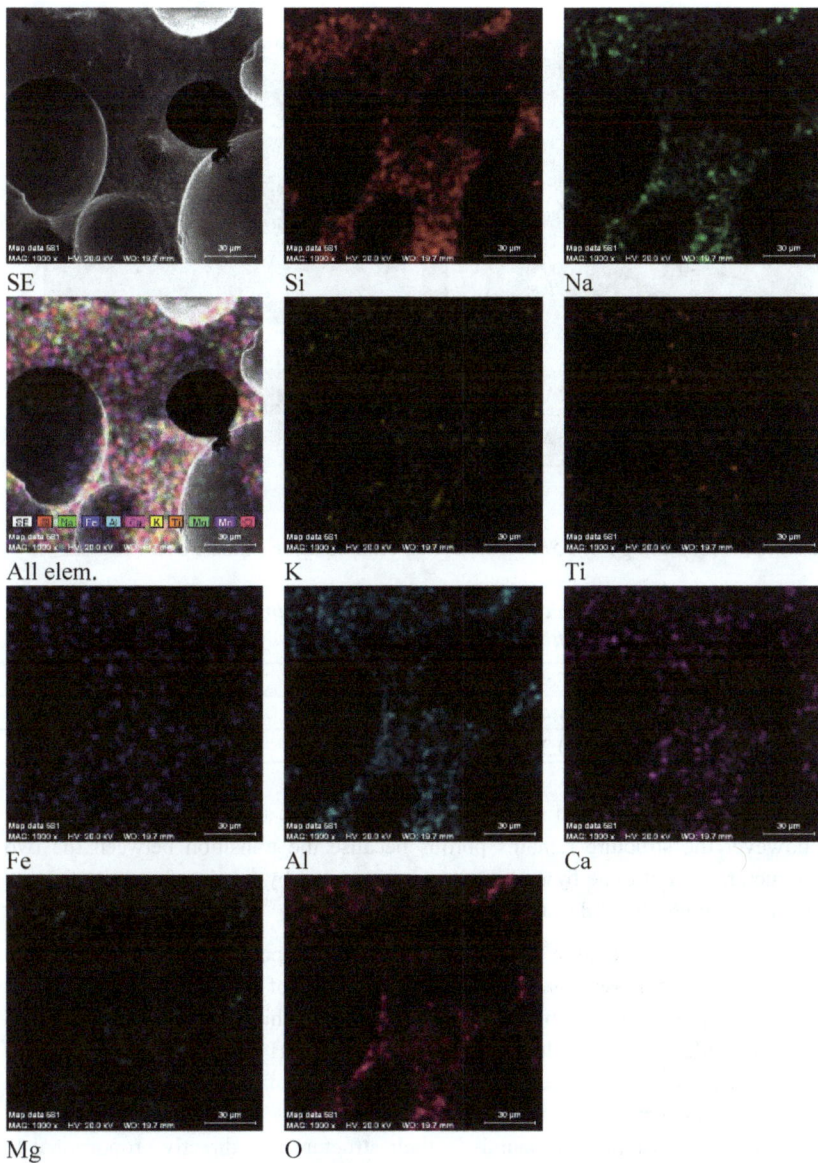

Figure 6.19. Elemental mapping of 30FA_70S sample.

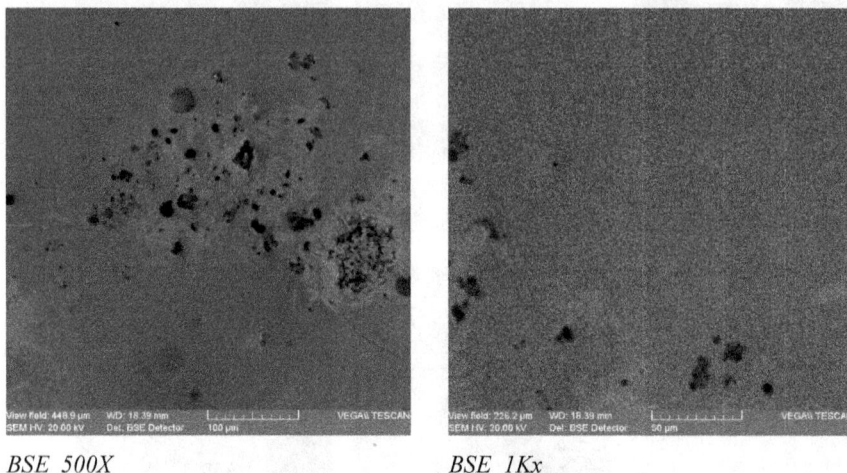

BSE_500X   BSE_1Kx

Figure 6.20. Morphology of 15FA_15PG_70S sample after exposure to 1000 °C.

Table 6.7. Chemical composition and elemental distribution in 15FA_15PG_70S sample after exposure to 1000 °C.

Element	Si	Na	Fe	Al	Ca	K	Ti	Mg	O
[%], wt.	28.41	10.94	3.46	3.89	5.16	1.95	1.11	1.09	balance
St. error, [%]	1.40	0.90	0.26	0.29	0.28	0.16	0.13	0.15	-

Exposure to high temperatures of the geopolymers produces a complete sintering of the matrix, however the structure remains porous because the transition between the two types of structures is not done by melting the skeleton, but by passing from one form to another of the main phases of the structure.

During the analyzed temperature range, the ash-based geopolymers based on thermal power plants show high mass losses, due to the elimination of free and physically bound water molecules up to 300 °C. Above this temperature, goethite, calcium hydroxide and aluminum hydroxide decompose due to the elimination of OH⁻ (chemically bound water) groups. However, it has been observed that the percentage of mass loss depends on the ash content of the power plant. Therefore, the hygroscopicity of the samples, as well as the concentration of unstable compounds in their structure, are directly proportional to the percentage of volume represented by the matrix.

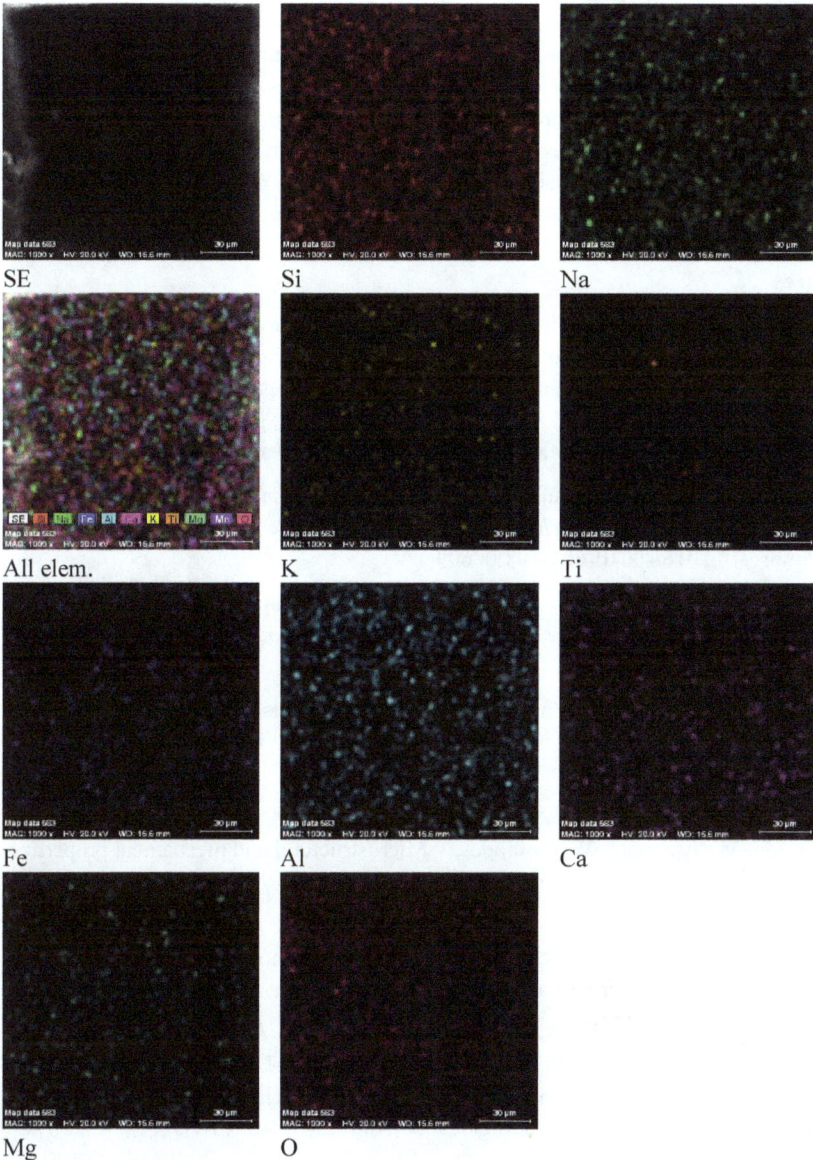

Figure 6.21. Elemental mapping of 15FA_15PG_70S sample.

Taking into account the large number of chemical elements and the possibilities of decomposition or formation of some compounds, it was found that during the heating there was a change in the initial phases of the geopolymers. By introducing the glass powder into the matrix, a denser sample was obtained, but, due to the high calcium content, the thermal stability decreases. Therefore, in order to obtain a geopolymer with high thermal stability, it is necessary to introduce as high a percentage of sand as possible in the matrix.

References

1.   Cornejo, M.H.; Togra, B.; Baykara, H.; Soriano, G.; Paredes, C.; Elsen, J. Effect of calcium hydroxide and water to solid ratio on compressive strength of mordenite-based geopolymer and the evaluation of its thermal transmission property. In Proceedings of the ASME International Mechanical Engineering Congress and Exposition, Proceedings (IMECE); American Society of Mechanical Engineers (ASME), 2018; Vol. 12.

2.   MacKenzie, K.J.D.; Temuujin, J.; Okada, K. Thermal decomposition of mechanically activated gibbsite. *Thermochim. Acta* 1999, *327*, 103–108. https://doi.org/10.1016/s0040-6031(98)00609-1

3.   Provencher, S.W. CONTIN: A general purpose constrained regularization program for inverting noisy linear algebraic and integral equations. *Comput. Phys. Commun.* 1982, *27*, 229–242. https://doi.org/10.1016/0010-4655(82)90174-6

4.   Lahoti, M.; Wong, K.K.; Yang, E.-H.; Tan, K.H. Effects of Si/Al molar ratio on strength endurance and volume stability of metakaolin geopolymers subject to elevated temperature. *Ceram. Int.* 2018, *44*, 5726–5734. https://doi.org/10.1016/j.ceramint.2017.12.226

5.   Criado, M.; Aperador, W.; Sobrados, I. Microstructural and mechanical properties of alkali activated Colombian raw materials. *Materials (Basel).* 2016, *9*. https://doi.org/10.3390/ma9030158

6.   Bede, A.; Scurtu, A.; Ardelean, I. NMR relaxation of molecules confined inside the cement paste pores under partially saturated conditions. *Cem. Concr. Res.* 2016, *89*, 56–62. https://doi.org/10.1016/j.cemconres.2016.07.012

7.   Pop, A.; Bede, A.; Dudescu, M.C.; Popa, F.; Ardelean, I. Monitoring the Influence of Aminosilane on Cement Hydration Via Low-field NMR Relaxometry. *Appl. Magn. Reson.* 2016, *47*, 191–199. https://doi.org/10.1007/s00723-015-0743-7

8.   Burduhos Nergis, D.D.; Vizureanu, P.; Ardelean, I.; Sandu, A.V.; Corbu, O.C.; Matei, E. Revealing the Influence of Microparticles on Geopolymers' Synthesis and

Porosity. *Materials (Basel).* 2020, *13*, 3211. https://doi.org/10.3390/ma13143211

9.   Lahoti, M.; Tan, K.H.; Yang, E.H. A critical review of geopolymer properties for structural fire-resistance applications. *Constr. Build. Mater.* 2019, *221*, 514–526.

10.   Temuujin, J.; Minjigmaa, A.; Rickard, W.; Lee, M.; Williams, I.; van Riessen, A. Preparation of metakaolin based geopolymer coatings on metal substrates as thermal barriers. *Appl. Clay Sci.* 2009, *46*, 265–270. https://doi.org/10.1016/j.clay.2009.08.015

11.   Van Dao, D.; Trinh, S.H.; Ly, H.B.; Pham, B.T. Prediction of compressive strength of geopolymer concrete using entirely steel slag aggregates: Novel hybrid artificial intelligence approaches. *Appl. Sci.* 2019, *9*. https://doi.org/10.3390/app9061113

12.   Hein, P.R.G.; Brancheriau, L. Comparison between three-point and four-point flexural tests to determine wood strength of Eucalyptus specimens. *Maderas Cienc. y Tecnol.* 2018, *20*, 333–342. https://doi.org/10.4067/S0718-221X2018005003401

13.   Wuddivira, M.N.; Robinson, D.A.; Lebron, I.; Bréchet, L.; Atwell, M.; De Caires, S.; Oatham, M.; Jones, S.B.; Abdu, H.; Verma, A.K.; et al. Estimation of soil clay content from hygroscopic water content measurements. *Soil Sci. Soc. Am. J.* 2012, *76*, 1529–1535. https://doi.org/10.2136/sssaj2012.0034

14.   Longhi, M.A.; Zhang, Z.; Rodríguez, E.D.; Kirchheim, A.P.; Wang, H. Efflorescence of alkali-activated cements (geopolymers) and the impacts on material structures: A critical analysis. *Front. Mater.* 2019, *6*. https://doi.org/10.3389/fmats.2019.00089.

15.   Liew, Y.-M.; Heah, C.-Y.; Mustafa, A.B.M.; Kamarudin, H. Structure and properties of clay-based geopolymer cements: A review. *Prog. Mater. Sci.* 2016, *83*, 595–629. https://doi.org/10.1016/j.pmatsci.2016.08.002

16.   Rico, P.; Adriano, A.; Soriano, G.; Duque, J. *V International Symposium on Energy Characterization of Water Absorption and Desorption properties of Natural Zeolites in Ecuador.*

17.   Valdiviés-Cruz, K.; Lam, A.; Zicovich-Wilson, C.M. Chemical interaction of water molecules with framework Al in acid zeolites: a periodic ab initio study on H-clinoptilolite. *Phys. Chem. Chem. Phys.* 2015, *17*, 23657–23666. https://doi.org/10.1039/c5cp03268g

18.   Calero, S.; Gómez-Álvarez, P. Hydrogen bonding of water confined in zeolites and their zeolitic imidazolate framework counterparts. *RSC Adv.* 2014, *4*, 29571–29580. https://doi.org/10.1039/c4ra01508h

19.   Glad, B.E.; Kriven, W.M. Geopolymer with Hydrogel Characteristics via Silane

Coupling Agent Additives. *J. Am. Ceram. Soc.* 2014, *97*, 295–302.
https://doi.org/10.1111/jace.12643

20.   Chindaprasirt, P.; Chareerat, T.; Hatanaka, S.; Cao, T. High-Strength Geopolymer Using Fine High-Calcium Fly Ash. *J. Mater. Civ. Eng.* 2011, *23*, 264–270
https://doi.org/10.1061/(asce)mt.1943-5533.0000161

21.   Palomo, A.; Krivenko, P.; Garcia-Lodeiro, I.; Kavalerova, E.; Maltseva, O.; Fernández-Jiménez, A. A review on alkaline activation: new analytical perspectives ; Activación alcalina: Revisión y nuevas perspectivas de análisis. 2014, *64*, 22.
https://doi.org/10.3989/mc.2014.00314

22.   Derie, R.; Ghodsi, M.; Calvo-Roche, C. DTA study of the dehydration of synthetic goethite αFeOOH. *J. Therm. Anal.* 1976, *9*, 435–440.
https://doi.org/10.1007/BF01909409

23.   Walter, D.; Buxbaum, G.; Laqua, W. The mechanism of the thermal transformation from goethite to hematite*. *J. Therm. Anal. Calorim.* 2001, *63*, 733–748.
https://doi.org/10.1023/A:1010187921227

24.   Cheng-Yong, H.; Yun-Ming, L.; Abdullah, M.M.A.B.; Hussin, K. Thermal Resistance Variations of Fly Ash Geopolymers: Foaming Responses. *Sci. Rep.* 2017, *7*.
https://doi.org/10.1038/srep45355

25.   Bajare, D.; Vitola, L.; Dembovska, L.; Bumanis, G. Waste Stream Porous Alkali Activated Materials for High Temperature Application. *Front. Mater.* 2019, *6*.
https://doi.org/10.3389/fmats.2019.00092

26.   Zhu, B.; Fang, B.; Li, X. Dehydration reactions and kinetic parameters of gibbsite. *Ceram. Int.* 2010, *36*, 2493–2498, doi:10.1016/j.ceramint.2010.07.007.

27.   Hao, H.; Lin, K.-L.; Wang, D.; Chao, S.-J.; Shiu, H.-S.; Cheng, T.-W.; Hwang, C.-L. *Elucidating characteristics of geopolymer with solar panel waste glass*; 2015; Vol. 14;.

28.   Grawe, S.; Augustin-Bauditz, S.; Clemen, H.C.; Ebert, M.; Eriksen Hammer, S.; Lubitz, J.; Reicher, N.; Rudich, Y.; Schneider, J.; Staacke, R.; et al. Coal fly ash: Linking immersion freezing behavior and physicochemical particle properties. *Atmos. Chem. Phys.* 2018, *18*, 13903–13923. https://doi.org/10.5194/acp-18-13903-2018

29.   Duan, P.; Yan, C.; Zhou, W. Compressive strength and microstructure of fly ash based geopolymer blended with silica fume under thermal cycle. *Cem. Concr. Compos.* 2017, *78*, 108–119. https://doi.org/10.1016/j.cemconcomp.2017.01.009

30.   Paunović, P.; Petrovski, A.; Načevski, G.; Grozdanov, A.; Marinkovski, M.;

Andonović, B.; Makreski, P.; Popovski, O.; Dimitrov, A. Pathways for the production of non-stoichiometric titanium oxides. In *Nanoscience Advances in CBRN Agents Detection, Information and Energy Security*; Springer Netherlands, 2015; pp. 239–253 ISBN 9789401796972.

31.   Anderson, P.J.; Horlock, R.F. Thermal decomposition of magnesium hydroxide. *Trans. Faraday Soc.* 1962, *58*, 1993–2004. https://doi.org/10.1039/tf9625801993

32.   Zulkifly, K.; Yong, H.C.; Abdullah, M.M.A.B.; Ming, L.Y.; Panias, D.; Sakkas, K. Review of Geopolymer Behaviour in Thermal Environment. In Proceedings of the IOP Conference Series: Materials Science and Engineering; Institute of Physics Publishing, 2017; Vol. 209.

33.   Gheorghe. T. Pop; Mihai Chiriță; Mihai Rostami *Materiale bioceramice*; Tehnopress: Iasi, 2003.

7. Sustainability with Geopolymers

7.1. Geopolymers technology for green cities

Sustainability, the goal of the 21st century, is defined in many ways, depending on the field in which it is applied, however, basically sustainability refers to the development of a particular material, object or field without producing negative effects on other areas, in economically, environmentally and socially terms. In the case of geopolymers, sustainable development refers to the production of a material that uses as raw materials sources that produce positive effects on the three factors involved. From an economic point of view, they have lower manufacturing costs than conventional materials (Figure 7.1), socially, the geopolymers concept strengthens the integration of the composite materials in social and industrial applications which enhancing and fostering collaboration between researcher and environmentally due to the fact that wastes can be used as raw material, consequently a lower embodied energy and carbon footprint at the proposed technology due to the processes at relatively low temperature.

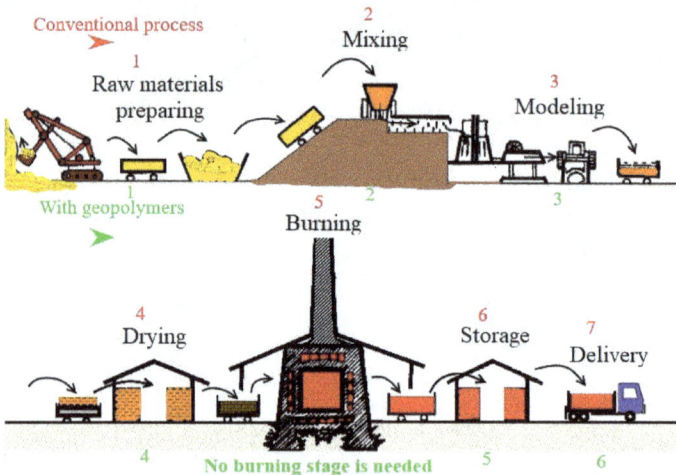

Figure 7.1. Process flow of conventional and geopolymeric bricks manufacturing [1].

Figure 7.1 shows the difference between the manufacturing process of conventional (clay based) blocks and bricks and those with geopolymers. The process of obtaining conventional blocks and bricks includes the following steps: first is the purchase/collecting of certified basic materials, then they must be transported to the factory, then their storage in silos, followed by metering and dry mixing, then mixing these materials with water. and their tuning in the press. These autoclaves combine pressure with vibration to obtain a high-performance prefabricated and measure particle size with interchangeable iron equipment. After obtaining the required granulation, the product must be dried for 24 hours at room temperature before being packaged and stored. By using geopolymers, multiple steps from the process flow can be eliminated and an ecofriendly product can be obtained.

Gradually city by city adopted the principle of "zero waste" by means of different recycling or recirculation methods. Although, this concept can be applied when we refer to plastic, cardboard or metal materials recycling, well, when is about oxidic waste, the possibilities of waste conversion in useful materials are very limited.

7.2. Environmental impact of geopolymers

The building materials sector is currently positioned on the third-place, in terms of CO_2 emissions, accounting about 10 % of the total CO_2 emissions globally, from this the largest amount is associated with the concrete manufacturing [2,3]. About 85 % of CO_2 emissions come from the production of cement, of which about 95 % is released during its manufacture and 5 % during the transport of raw materials and finished products [4].

Over the last year, multiple studies have been conducted on the environmental impact of the construction and building materials. Generally, those studies evaluate the effects produced on environment (CO_2 emission, raw materials consumption etc.) by any stage of manufacturing process of cement. In the case of geopolymers, the same method has been applied, therefore, the overall environmental impact of geopolymers must be calculated considering the production of components (sodium hydroxide or silicate, potassium hydroxide or silicate, lithium activator, raw materials extraction or collection, raw materials calcination or drying, transportation of the components, mixing, curing, packaging, final products delivery etc.). In other words, to correctly and accurately evaluate the environmental impact of geopolymers, it must be considered all the components involved in the product obtaining and all the manufacturing stages, i.e. from components production up to placement on site and further recycling (Figure 7.2).

G. Habert et al. [5] focused their research on analyzing the environmental impacts of alkali activated materials associated with the production of the constituents. According to their research the geopolymers used as substitution materials for cement contributes to the reduction of the CO_2 emissions by a factor of 4. However, in terms of the impact related to less critical construction industry (raw materials consumption etc.), the geopolymers seems to remain on the same level as Ordinary Portland Cement (OPC) materials, especially, in the case of kaolin based geopolymers.

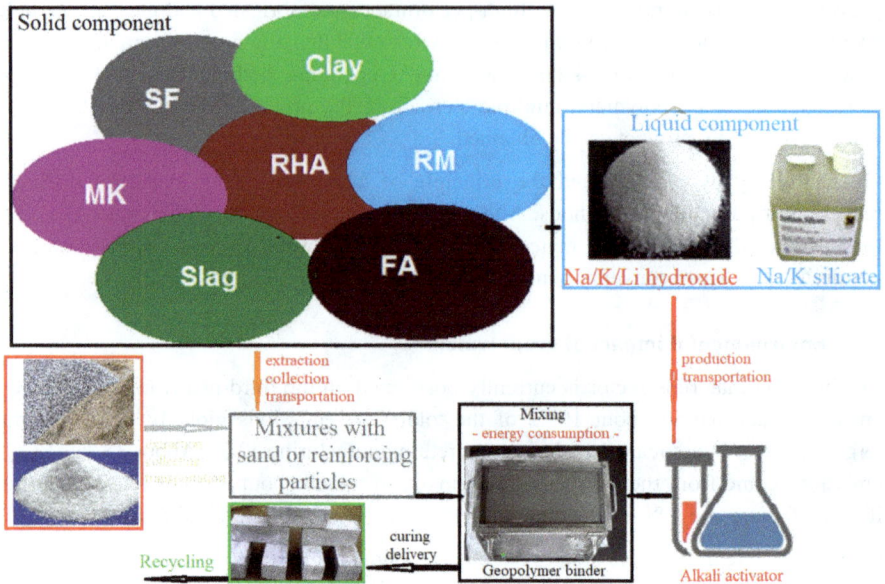

Figure 7.2. Factors that must be considered during environmental impact evaluation.

In the study conducted by R. Bajpai et al. [6] on the environmental impact of fly ash based geopolymers with or without addition of silica fume it has been concluded that geopolymers are more affected by the raw materials transportation than cement concrete. Also, the freshwater ecotoxicity and human toxicity of the studied mixes are higher in the case of geopolymers comparing to the values obtained for the OPC materials. Besides, the main component (responsible for 59.97%) affecting the environmental impact is the alkaline activator. However, the overall endpoint score on ecosystem, human health, and resources of the studied geopolymers are less than cement concrete, providing better performance in terms of climate change and sustainability.

A comparison between geopolymers and OPC concrete in terms of carbon dioxide equivalent (CO_2-e) emissions has been presented in L. K. Turner et al. [7]. According to their study the CO_2 footprint of geopolymer concrete was approximately 9% lower than that of OPC considering the impact produced by the factors presented in Figure 7.3.

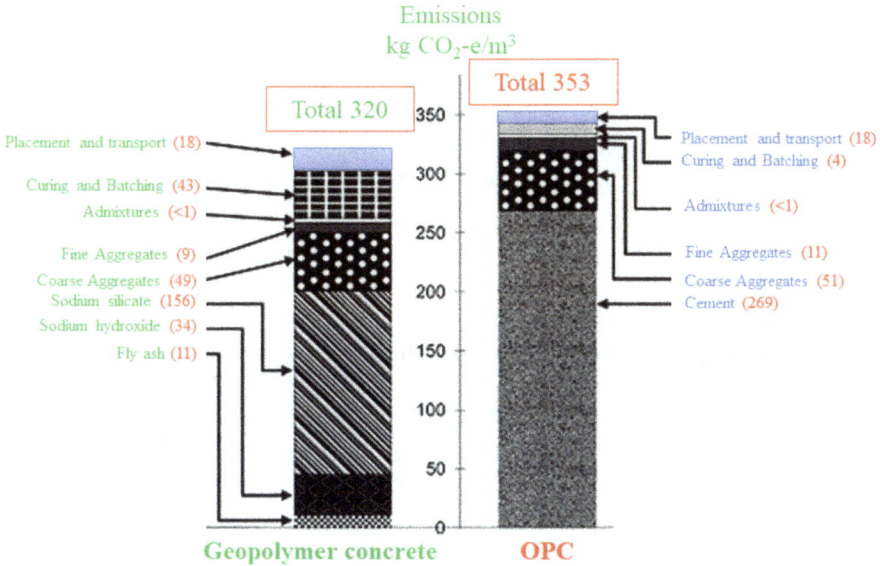

Figure 7.3. CO_2-e for fly ash based geopolymers and OPC [8].

Moreover, the development of geopolymers was also encouraged by the lower amount of CO_2 emissions in their case, compared to those resulting from the manufacture of Portland cement concrete [9–12]. However, several studies negatively assess [13,14] this aspect of CO_2 emissions, but according to the study published by J. Davidovits [10], those studies are based on erroneous assessment methodologies or miscalculations. According to J. Davidovits studies, the greenhouse gas emissions of geopolymers are 62 % - 66 % lower than those resulting from the manufacture of Portland cement concrete.

Based on the principles of the process intensification strategy it can lead to the development and the design of new processes more compact and efficient that allow the obtaining of advanced materials with lower energy consumption and better exploitation of raw materials (use waste as raw materials).

7.3. Applications

Due to their low cost and high mechanical and chemical properties, these oxide materials have been of particular interest for the manufacture of thermal insulation, fire-resistant materials, construction materials, decorative objects, refractory linings, as well as for the repair and consolidation of infrastructures or encapsulation of radioactive and toxic waste, etc. [15]. However, the main application for geopolymers has been identified in the construction industry as environmentally friendly concrete, with low energy consumption and low CO_2 footprint, compared to conventional Portland cement concrete.

By analogy with petroleum-derived organic polymers, geopolymers are polymers synthesized by polycondensation and hardening into inorganic materials with mechanical properties comparable to those of conventional materials. These properties have given a very strong push to the creativity and innovation of geopolymer researchers in the development of new compositions and the identification of possible applications [16].

7.3.1. Civil engineering

Currently, geopolymers are used in a variety of applications in the field of civil engineering [17] for the construction of buildings where they are used in the manufacture of bricks, facade panels, interior walls, floors etc. (Figure 7.4). The choice of material for each product is made according to its characteristics. In the case of elements that are part of the resistance structure (columns, beams) the most important property is the compressive strength. In the case of insulation elements (panels, walls or bricks) thermal conductivity and fire resistance are the most important characteristics [18]. However, in the case of surface elements (facades, floors, etc.) roughness, homogeneity and low density are a priority [19].

The geopolymers were also swiftly introduced into the market as road infrastructure materials, especially for the manufacture of paving bricks, covering pipes, sidewalks etc. [20].

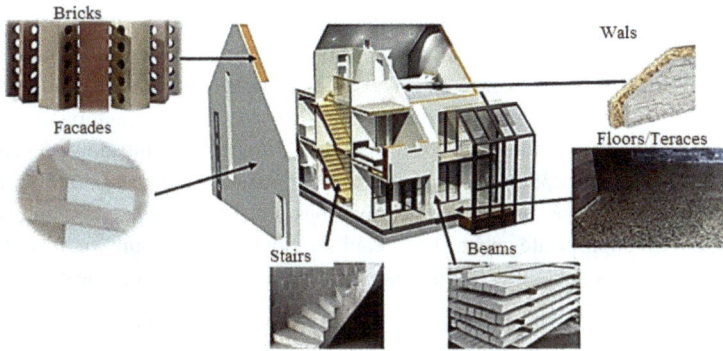

Figure 7.4. Successfully applications of geopolymers in civil building components.

7.3.2. Geopolymers as multifunctional materials

After the successful implementation of geopolymers as building materials, researchers began to develop them for applications that required high operating temperatures. Therefore, since the 1990s they have been used in the manufacture of elements surrounding the exhaust system of Formula 1 cars, replacing the titanium components [19], as well as in the manufacture of filters or refractory liners (Figure 7.5).

Around the 2000s they were introduced in the manufacture of the inner shell of aircraft, due to its low density, surface characteristics and excellent fire resistance [19].

a)   b)   c)

Figure 7.5. Other applications of geopolymers. a) refractory linings; b) open-pores structures (filters); c) high temperature resistance coatings.

The multifunctional geopolymers are created through the chemical reaction which occurs after mixing the raw materials with an alkaline solution. This multiple-stage reaction consists of: (1) dissolution of the aluminosilicate source in alkaline medium, (2) Si-O-Al network and gel formation and (3) formation of geopolymer structure. In case of water purification, knowing that this process involves the separation of a matter (microparticles, heavy metals etc.) from one phase (aqueous solution) and its accumulation at the surface of another phase, the resulted material will be designed for ultralow density (high porosity) and super-adsorption capacity. In order to simulate the best condition in which the materials could be used as an adsorbent, various physico-chemical parameters such as the selection of appropriate electrolyte, equilibration time, amount of adsorbent, the concentration of adsorbate, effect of diverse ions and temperature must be studied.

In case of geopolymers, through an appropriate process technology a self-supporting membrane with tailored properties can be obtained which is expected to offer outstanding opportunities for energy efficient separations and process intensification, in terms of saving energy, reducing capital costs, minimizing environmental impact and maximizing the raw materials exploitation by using wastes (fly ash, red mud, slags etc.) as raw materials.

Up to know it has been proven that, geopolymer membranes offer significant advantages over polymeric or metallic membranes in many applications with extreme operating conditions because of their intrinsic properties i.e., rigid porous structure, high-temperature resistance, high-chemical resistance to aggressive aqueous and organic media, insensitiveness to biological attack. Therefore, this type of geopolymers are suitable for: the treatment of waste liquids and gases; liquid processing including drinking water, domestic water, and food beverages; and product recovery in various industries range from micrometer-sized species (mineral particles, microorganisms, macromolecules, etc.), nanometer-sized species (viruses, colloids, molecules, ions), to filtration pretreatment before other separation techniques including polymer membranes.

In the beginnings, the geopolymers were used as building materials for the manufacture of building panels with high fire resistance [27]. Subsequently, due to its advantages compared to conventional materials, geopolymers were introduced in the automotive, ceramics, metallurgy, aerospace etc. industries. According to the literature, any material, in powder form, with a high content of aluminum and silicon oxides can be used as a raw material [20,32] for geopolymers, however, the method of production must be customized according to the desired final characteristics.

7.3.3. Smart self-sensing and monitoring geopolymers

Due to their high conductibility, compatibility with construction materials and sensing capacity, the geopolymers can be used in the construction of smart self-sensing and monitoring devices. Y. Wu et al. [21] described the main characteristics that promote the geopolymers for the manufacturing of this type of devices. In their study a physical-chemical model based on the imbalance of charge and the migration of dipoles generated under external pressure, describes the direct piezoelectric effect of geopolymers. According to C. Lamuta et al. [22] the metakaolin based geopolymers can be used for the manufacturing of devices for real time self-monitoring of civil infrastructures.

The self-sensing performance of geopolymers has also been studied in S. Bi et al. [23], according to this research the geopolymers reinforced with carbon nanotubes exhibit ultrahigh self-sensing performance based on the unique behaviors of SiO_2 coating on carbon nanotubes in the matrix. This behaviour is mostly related with the fact that the layer of SiO_2 from the carbon nanotubes was partially or fully removed during the fabrication process to restore the conductive nature of carbon nanotubes, facilitating the dispersion of carbon nanotubes and forming well-connected 3D electrical conductive networks. Moreover, according to their study the gauge factor of geopolymer nanocomposites reached up to 663.3 and 724.6, under compressive and flexural loading, respectively, with the addition of only 0.25 vol % of SiO_2-coated carbon nanotubes.

7.3.4. Biomedical materials

The oxidic inorganic materials used in the biomedical applications field are mainly divided into two categories: bioinert materials and bioactive materials [21]. Due to the fact that geopolymers can be shaped, cured at room temperature and promote bone growth and recovery, different researchers studied the possibility to use those composites as bioinert materials for joints manufacturing [24]. However, the use of geopolymers as biomedical materials is limited by their know alkalinity, moreover, a low concentration of aluminium stimulates the new bone formation and proliferation of osteoblasts, but, in the same time, aluminium presence may prove toxic to biological cells when implanted in living organisms.

According to the H. Oudadesse et al. [25] the main limitation, alkalinity, can be solved by immobilizing the functional groups through a heat-treatment at high temperature which will reduce the pH from approximatively 11 to 7. Moreover, due to the non-toxicity of metakaolin based geopolymers, their biocompatibility and drug carrying capacity, those materials can be used as oral drug delivery carrier.

7.4. Short overview on the geopolymers engineering applications

Geopolymers are inorganic materials formed by the geopolymerization reaction that occurs between a material based on silicon and aluminum oxides and an alkaline solution.

Their structure is based on tetrahedral Si-O-Al (sialate) bonds that are chemically balanced by group I alkaline ions (Na^+, K^+ or Li^+) [26].

In the begining, the geopolymers were mainly used as building materials for the manufacture of sawdust panels with high fire resistance. Subsequently, due to the advantages they present, compared to conventional materials, geopolymers were introduced in the automotive, ceramics, metallurgy and aerospace industries. According to the literature, any material, in powder form, with a high content of aluminum and silicon oxides can be used as a raw material for geopolymers, however, the production method must be customized according to the desired final characteristics (tailored properties). Considering that the self-supporting geopolymeric membranes could have tailored properties, including superabsorbent capacity, heavy metals encapsulation, flame retardancy, mechanical performance, electrokinetic behaviour, corrosion resistance and thermal properties, due to their porous inorganic matrix [27,28].

The critical analysis of the literature reveals that, despite the many advantages, worldwide there are still many challenges to be overcome for the successful introduction in industrial applicatiosn of geopolymers. These are related to the lack of long-term evaluations , in real environmental conditions, and standards for geopolymers. Moreover, geopolymers that use certain natural minerals (kaolin, clay etc.) as a source of raw materials are expensive and produce negative effects on mining areas [29–31]. By comparison, the presented research addresses the obtaining of geopolymers, without Portland cement, by recovering some mineral waste (thermal power plant ash and glass powder), thus producing positive effects on the environment, by greening storage areas or dumps.

Although worldwide, geopolymers are of great interest for research in the field of materials engineering, in Romania the concerns for these materials are quite low. However, at national level, there are large amounts of mineral waste (power plant ash, red sludge, mining tailings, etc.) with potential for geopolymerization that could be exploited through this technology [32,33].

Considering the diversity of raw material sources, the different methods of obtaining and the parameters with major influence on the properties of geopolymers, presented in the literature, it was necessary to evaluate and experimentally determine the optimal parameters for the process. Thus, the study of the following parameters was established: the characteristics of the raw material, the ratio between the alkaline activator components, the ratio between solid and liquid, the curing time, the curing temperature and the possibility of adding reinforcing particles [27,34].

The experimental research program aimed at the design, obtaining / elaboration and characterization of geopolymers based on indigenous thermal power plant ash, considering a complex and interdisciplinary study in the field of physics, chemistry, materials science and civil engineering on oxide materials. based on silicon and aluminum.

As a result of the study of the factors influencing the obtaining of these materials and of the potential sources of local raw material, four types of geopolymers based on thermal power plant ash and reinforcement particles were designed.

To obtain the four types of materials, the following was used: (i) power plant ash, with particles smaller than 80 µm, (ii) glass powder, with particles smaller than 10 µm and (iii) sand with particles smaller than 4 m mm. The samples were dried at a temperature of 70 °C for three different periods of time: 8 hours, 16 hours, respectively 24 hours. Alkaline activation was performed using a solution of sodium silicate and sodium hydroxide with a molar concentration of 10 at a mass ratio of 1.5. The ratio between the solid component and the activation solution, determined experimentally, is 1. The solid component of the first type of geopolymer (sample 100FA) contains 100% ash from the power plant, the second type (70FA_30PG) is made from 70% ash from thermal power plant and 30% glass powder, the third (30FA_70S) contains 30% thermal power plant ash and 70% sand, and the fourth type (15FA_15PG_70S) is made of 15% thermal power plant ash, 15% powder glass and 70% sand.

The production of these new geopolymers based on thermal power plant ash is a method of capitalization of these mineral wastes, in order to obtain new materials with mechanical and chemical properties, comparable to those of conventional materials (based on Portland cement) [35].

In order to evaluate the main characteristics of the obtained geopolymers, it was necessary to evaluate the samples from the chemical (chemical composition), structural (macrostructural, microstructural and mineralogical), physical-mechanical (setting time, relative pore distribution, compressive strength and tensile strength). bending) and thermal (thermogravimetric analysis and differential thermal analysis).

In order to improve the mechanical properties of the obtained geopolymers, the effect of introducing, as a reinforcing element, the glass particles, which come from the recycling of some wastes from the food industry, was studied.

The novelty of the research is based on (i) the originality of the chemical composition of the raw material and the reinforcing elements and (ii) on the technology of obtaining geopolymers. The developed materials are obtained with minimum energy consumption and optimized in terms of mechanical properties through the activation solution and the percentage of reinforcing particles.

In order to fulfill the main objective of the presented research, which refers to obtaining geopolymers by capitalizing on mineral waste, several stages of experimental research have been established:

- identification, sampling and characterization of raw materials;
- design of geopolymers as environmentally friendly materials with appropriate properties for industrial applications;
- obtaining four types of geopolymers based on thermal power plant ash with or without reinforcement particles;
- characterization of geopolymers obtained from the chemical, structural, mechanical and thermal behavior;
- evaluation and identification of geopolymers with the best chemical, structural, mechanical and thermal characteristics.

In the first stage, several preliminary experiments were carried out which led to the identification of raw materials: power plant ash, glass powder and sand. Subsequently, the sources of mineral waste, the ash from the thermal power plant (the storage dumps of the company CET II- Holboca, Iași), as well as the glass powder (collected by RECYCLE International and ground by New NCR Reciclare SRL) were identified. , respectively of the sand (class $(0 \div 4)$ mm, provided by Ungureanu Trans SRL). Then, the raw materials taken were characterized, the results highlighting the fact that the raw materials can be used to obtain geopolymers (has the potential for geopolymerization).

The materials design, in the form of the four types of geopolymers, had as main criterion the compressive strength of geopolymers. Therefore, the proportion and type of reinforcing elements (glass powder or sand) was established according to their influence on the mechanical properties of geopolymers.

To obtain the geopolymer samples, a production method was designed which includes the following steps: preparation of the liquid and solid components, mixing of the components, obtaining the geopolymer paste, pouring the paste into shapes and drying / obtaining the actual four types of geopolymers.

The analysis of the chemical composition of the samples highlights the differences produced by the reduction of the amount of power plant ash by replacing it with reinforcing particles. Sample 100FA contains the highest concentration of Fe and Al, but the lowest of Si. Sample 70FA_30PG shows a decrease of more than 20% in the concentration of Fe and Al, but due to the replacement of 30% of the ash from the glass powder plant, the calcium concentration in the sample increases by approximately 30%. The chemical composition of the 30FA_70S sample shows an increase of approximately 17% in the Si concentration, but, by replacing 70% of the amount of power plant ash with sand particles, the Al and Ca content decreases by 46% and 3%, respectively. In the case of the 15FA_15PG_70S sample, the increase of the Si concentration is approximately 17%, and the decrease of the aluminum content reaches 50%.

From a structural point of view, the geopolymer samples obtained have a homogeneous chemical distribution. However, from a macroscopic point of view, a different layer with a thickness of about 3 mm is observed in the upper part of the samples. Its morphology suggests a separation, during the vibration stage, of impurities based on differences in density.

The microstructural analysis of the 100FA sample reveals a large number of unreacted ash particles throughout the volume, as well as cracks formed during the drying process as a result of the rapid evaporation of water. After the introduction of the reinforcing particles into the structure (samples 70FA_30PG, 30FA_70S and 15FA_15PG_70S), a decrease in the number of unreacted particles can be observed, however, there is a significant increase in the number of large pores (> 600 nm) due to blockage of air bubbles. by aggregates.

From a mineralogical point of view, the ash of the thermal power plant used contains mainly oxide compounds of aluminum and silicon. Most of the maxima present on the diffractograms are positioned between 20 ° and 45 ° (2θ), and the phase corresponding to the maximum with the highest intensity is quartz (26.62 °, 2θ). Following activation, an additional phase specific to zeolites appears, sodalite (24.5 °, 2θ), which confirms that the activation of the geopolymer has taken place.

The analysis by IR spectroscopy with the Fourier transform of the obtained geopolymer samples highlights their chemical structure based mainly on groups formed between silicon, oxygen and aluminum atoms, but also hydrogen. After activation, in addition to the characteristic gray power plant strips, glass particles or sand, there are also sialate-specific strips (Si-O-Al) that confirm the geopolymerization reaction between the raw material and the activator, as well as a specific band of water molecules, tape that highlights the hygroscopic character of these oxide materials.

The FTIR spectrum of the power plant ash and the glass powder used shows a wide vibration band between 3700 cm-1 and 3000 cm-1 which is assigned to the OH⁻ groups. The large bandwidth is due to the high degree of association of hydrogen with other hydroxyl groups by creating strong bonds between the OH⁻ and Si ($\equiv$Si – OH) groups.

The second significant peak, between 1150 cm-1 and 1250 cm-1, can be associated with the band specific to the rhythmic motion along the axis of the covalent bond, which is known as elongation vibration, of the asymmetric group of Si-O -And. The vibration band between 800 cm-1 and 700 cm-1 is specific to the asymmetric groups of Si-O-Al in the compounds of the analyzed material, and the band between 700 cm-1 and 600 cm-1 is assigned to the ring tetrahedra of Si-O .

Geopolymer samples based on thermal power plant ash (dried / hardened for 8 hours) show several inflection points on the graphs of the relative distribution of pores obtained by NMR. The first point between ≈0.1 ms and 1 ms corresponds to the gel-like pore fluid (<50 nm). The second point between ≈ 1 ms and 7.5 ms corresponds to the capillary pores (50 ÷ 600) nm, and the third point> 7.5 ms corresponds to the liquid in the gaps (pores larger than 600 nm) or the resulting cracks following the increase of the crystalline phases.

After studying the influence of drying time and the percentage of power plant ash on the relative distribution of pores in the structure of geopolymers, it was found:

- the increase of the percentage of reinforcing particles in the composition leads to the decrease of the relaxation times of the gel-type pores;
- the increase of the drying period causes the decrease of the relaxation times from the gel-type pores, respectively the increase of the relaxation times specific to the capillary pores or to the large pores.

The obtained geopolymers show the beginning of setting after (4.78 ÷ 6.88) hours and the end of setting after (22.95 ÷ 25.00) hours, at ambient temperature. According to the obtained results, by introducing the glass powder in the composition of the geopolymers there is a delay (increase of time) of the specific points of the setting time. However, the samples with sand particles in the composition showed an advance of the starting point and the end point of the setting.

From the point of view of compressive and bending strength, the elaborated ash-based geopolymers are positively influenced by the introduction of reinforcing particles. In the case of compressive strength, the value increases by up to 60%, and in the case of bending strength the difference reaches 15% in the case of samples aged 90 days.

During the compressive stress of the samples, the cracks formed advance through the interface between the matrix and the reinforcing elements, thus respecting the law of lowest resistance. The microstructural analysis of the samples highlights their cracking mechanism, observing that the propagation of cracks takes place after a sinusoidal trajectory through the interface between the matrix and the reinforcing elements, but also through the middle of the pores.

The decomposition of calcium hydroxide and goethite during heating was identified by TG-DTA analysis and confirmed by XRD analysis on samples heated to 1000 ± 5 ° C, respectively. The diffractograms of these samples show a decrease in the intensity of the maximum points specific to calcium hydroxide and goethite, as well as a sharp increase in hematite and calcium carbonate.

In the analyzed temperature range $(25 \div 340)$ ° C, the geopolymer samples show an inflection point, specific to the evaporation of free or physically bound water from the composition. Evaporated water highlights the hygroscopic nature of geopolymers and can be correlated with the amount of water detected by NMR in C-S-H, C-A-S-H and N-A-S-H pores in the structure of samples kept in the air for 48 days. The hygroscopic character of geopolymers was also confirmed by FTIR analysis by detecting the plane deformation vibration of the angle formed between hydrogen and oxygen atoms in the δ-H-O – H bond, specific to adsorbed water molecules.

During the analyzed temperature range, the ash-based geopolymers based on thermal power plants show high mass losses, due to the elimination of free and physically bound water molecules up to 300 ° C. Above this temperature, goethite, calcium hydroxide and aluminum hydroxide decompose due to the removal of OH^- (chemically bound water) groups. However, it has been observed that the percentage of mass loss depends on the ash content of the power plant. Therefore, the hygroscopicity of the samples, as well as the concentration of unstable compounds in their structure, are directly proportional to the percentage of volume represented by the matrix.

Taking into account the large number of chemical elements and the possibilities of decomposition or formation of some compounds, it was found that during the heating there was a change in the initial phases of the geopolymers. By introducing the glass powder into the matrix, a denser sample was obtained, but, due to the high calcium content, the thermal stability decreases. Therefore, in order to obtain a geopolymer with high thermal stability, it is necessary to introduce as high a percentage of sand as possible in the matrix.

As a result of the investigations carried out during the research and respecting the involved methodology, provided in Chapter 2, is presented a centralization of the experimental results for the four types of geopolymers obtained (Table 7.1). At the same time, we highlighted, where appropriate, the favorable behavior as environmentally friendly materials with appropriate properties for industrial applications and corresponding to the legal norms in force (ASTM C109 / C109M-07 Standard Test Method for Compressive Strength of Hydraulic Cement Mortars (Using 2- in or [50-mm] Cube Specimens) SR EN 197-1: Cement - Part 1: Composition, specifications and conformity criteria of common cements, ASTM C 191-04a Vicat Needle; SR EN 196-1 Cement testing methods).

Table 7.1. Synthesis of the specific characteristics of the obtained and characterized geopolymers

Property		Geopolymer sample			
		100FA	**70FA_30PG**	**30FA_70S**	**15FA_15PG_70S**
Phase composition		Sodalite, quartz, corundum, anorthite, hematite, portlandite, goethite, carbonate and albite.			
Mineral composition		OH⁻ and Si groups (≡Si - OH), asymmetric Si-O-Si groups, asymmetric Si-O-Al groups and Si-O ring tetrahedra.			
Setting time		Initial setting time (4.78 ÷ 6.88) h; Final setting time (22.95 ÷ 25.00) h.			
Porosity		Three categories of pores: gel type (<50 nm). capillary pores (50 - 600 nm). voids (pores greater than 600 nm).			
		4*	3*	2*	**1***
Compressive strength [MPa]		15.6	12.7	**27.2**	16.2
Flexural strength [MPa]		**10.2**	8.1	9.6	9.3
Thermal behaviour		Thermally stable between (25 ÷ 340) ° C; between (340 ÷ 1000) ° C shows the decomposition of several compounds (goethite, portlandite, etc.)			
Thermal stability, [%, wt.]	(25 ÷ 340) °C	23	19	12	**2**
	(340 ÷ 800) °C	28	21	13	**3**

*, synthetic coefficient; where: 1 is the sample with the lowest gel pore content and 4 is the sample with the highest gel pore content.

The developed geopolymer 100FA, 70FA_30PG and 15FA_15PG_70S belong, according to the standard NE 012-99, to concrete class C8/10 (used for leveling layers pouring in foundations), because they present compressive strength higher than 10 MPa at 28 days. At the same time, because the geopolymer 30FA_70S posses a compressive strength higher than 20 MPa at 28 days, this type of geopolymer belongs to class C16/20 and can be used for the execution of resistance structures of some industrial constructions. Moreover, due to the size (<16 mm) of the particles in the composition, all types of geopolymers obtained / elaborated can be used as pumping and plastering masses.

References

1.   Fabricarea caramizilor din argila: Rombadconstruct Available online: https://www.rombadconstruct.ro/fabricarea-caramizilor-din-argila.html (accessed on Aug 11, 2020).

2.   Andrew, R.M. Global CO 2 emissions from cement production. *Earth Syst. Sci.*

Data 2018, *10*, 195–217. https://doi.org/10.5194/essd-10-195-2018

3.   Tae, H.K.; Chang, U.C.; Gil Hwan, K.; Hyoung Jae, J. Analysis of CO2 Emission Characteristics of Concrete Used at Construction Sites. *Sustainability* 2016, *8*.

4.   Turner, L.K.; Collins, F.G. Carbon dioxide equivalent (CO2-e) emissions: A comparison between geopolymer and OPC cement concrete. *Constr. Build. Mater.* 2013, *43*, 125–130. https://doi.org/10.1016/j.conbuildmat.2013.01.023

5.   Habert, G.; d'Espinose de Lacaillerie, J.B.; Roussel, N. An environmental evaluation of geopolymer based concrete production: reviewing current research trends. *J. Clean. Prod.* 2011, *19*, 1229–1238. https://doi.org/10.1016/j.jclepro.2011.03.012

6.   Bajpai, R.; Choudhary, K.; Srivastava, A.; Sangwan, K.S.; Singh, M. Environmental impact assessment of fly ash and silica fume based geopolymer concrete. *J. Clean. Prod.* 2020, *254*, 120147. https://doi.org/10.1016/j.jclepro.2020.120147

7.   Turner, L.K.; Collins, F.G. Carbon dioxide equivalent (CO2-e) emissions: A comparison between geopolymer and OPC cement concrete. *Constr. Build. Mater.* 2013, *43*, 125–130. https://doi.org/10.1016/j.conbuildmat.2013.01.023

8.   Singh, N.B.; Middendorf, B. Geopolymers as an alternative to Portland cement: An overview. *Constr. Build. Mater.* 2020, *237*, 117455. https://doi.org/10.1016/j.conbuildmat.2019.117455

9.   Cheng, H.; Zhou, Y.; Liu, Q. Kaolinite Nanomaterials: Preparation, Properties and Functional Applications. In *Nanomaterials from Clay Minerals*; Elsevier, 2019; pp. 285–334.

10.   Davidovits, J. *Geopolymer Institute Library. Technical paper #24, False-CO2-values*; 2015.

11.   McLellan, B.C.; Williams, R.P.; Lay, J.; Van Riessen, A.; Corder, G.D. Costs and carbon emissions for geopolymer pastes in comparison to ordinary portland cement. *J. Clean. Prod.* 2011, *19*, 1080–1090. https://doi.org/10.1016/j.jclepro.2011.02.010

12.   Habert, G.; D'Espinose De Lacaillerie, J.B.; Roussel, N. An environmental evaluation of geopolymer based concrete production: Reviewing current research trends. *J. Clean. Prod.* 2011, *19*, 1229–1238. https://doi.org/10.1016/j.jclepro.2011.03.012

13.   Bajare, D.; Vitola, L.; Dembovska, L.; Bumanis, G. Waste Stream Porous Alkali Activated Materials for High Temperature Application. *Front. Mater.* 2019, *6*. https://doi.org/10.3389/fmats.2019.00092

14.   Olabemiwo, F.A.; Tawabini, B.S.; Patel, F.; Oyehan, T.A.; Khaled, M.; Laoui, T.

Cadmium Removal from Contaminated Water Using Polyelectrolyte-Coated Industrial Waste Fly Ash. *Bioinorg. Chem. Appl.* 2017, *2017*. https://doi.org/10.1155/2017/7298351

15. Singh, B.; Ishwarya, G.; Gupta, M.; Bhattacharyya, S.K. Geopolymer concrete: A review of some recent developments. *Constr. Build. Mater.* 2015, *85*, 78–90.

16. Davidovits, J. *Properties of Geopolymer Cements*; 1994;

17. Belviso, C. State-of-the-art applications of fly ash from coal and biomass: A focus on zeolite synthesis processes and issues. *Prog. Energy Combust. Sci.* 2018, *65*, 109–135.

18. Majidi, B. Geopolymer technology, from fundamentals to advanced applications: A review. *Mater. Technol.* 2009, *24*, 79–87. https://doi.org/10.1179/175355509X449355

19. Davidovits, J. *30 Years of Successes and Failures in Geopolymer Applications. Market Trends and Potential Breakthroughs.*

20. Safe and sustainable geopolymer concrete | Result In Brief | CORDIS | European Commission Available online: https://cordis.europa.eu/article/id/239529-safe-and-sustainable-geopolymer-concrete (accessed on Feb 16, 2020).

21. Wu, Y.; Lu, B.; Bai, T.; Wang, H.; Du, F.; Zhang, Y.; Cai, L.; Jiang, C.; Wang, W. Geopolymer, green alkali activated cementitious material: Synthesis, applications and challenges. *Constr. Build. Mater.* 2019, *224*, 930–949.

22. Lamuta, C.; Candamano, S.; Crea, F.; Pagnotta, L. Direct piezoelectric effect in geopolymeric mortars. *Mater. Des.* 2016, *107*, 57–64, doi:10.1016/j.matdes.2016.05.108.

23. Hu, S.; Wang, H.; Zhang, G.; Ding, Q. Bonding and abrasion resistance of geopolymeric repair material made with steel slag. *Cem. Concr. Compos.* 2008, *30*, 239–244. https://doi.org/10.1016/j.cemconcomp.2007.04.004

24. Yap, A.U.J.; Pek, Y.S.; Kumar, R.A.; Cheang, P.; Khor, K.A. Experimental studies on a new bioactive material: HAIonomer cements. *Biomaterials* 2002, *23*, 955–962. https://doi.org/10.1016/S0142-9612(01)00208-3

25. Oudadesse, H.; Derrien, A.C.; Lefloch, M.; Davidovits, J. MAS-NMR studies of geopolymers heat-treated for applications in biomaterials field. *J. Mater. Sci.* 2007, *42*, 3092–3098. https://doi.org/10.1007/s10853-006-0524-7

26. Nergis, D.D.B.; Abdullah, M.M.A.B.; Vizureanu, P.; Tahir, M.F.M. Geopolymers and Their Uses: Review. *IOP Conf. Ser. Mater. Sci. Eng.* 2018, *374*, 12019. https://doi.org/10.1088/1757-899x/374/1/012019

27. Burduhos Nergis, D.D.; Vizureanu, P.; Ardelean, I.; Sandu, A.V.; Corbu, O.C.; Matei, E. Revealing the Influence of Microparticles on Geopolymers' Synthesis and

Porosity. *Materials (Basel).* 2020, *13*, 3211. https://doi.org/10.3390/ma13143211

28.   Doru Dumitru, N.B.; Al Bakri Abdullah, M.M.; Petrică, V. The effect of fly ash/alkaline activator ratio in class F fly ash based geopolymers. *Eur. J. Mater. Sci. Eng.* 2017, *2*, 111–118.

29.   Abdullah, M.M.A.B.; Ming, L.Y.; Yong, H.C.; Tahir, M.F.M. Clay-Based Materials in Geopolymer Technology. In *Cement Based Materials*; InTech, 2018.

30.   Davidovits, J. Geopolymers Based on Natural and Synthetic Metakaolin a Critical Review. In; 2018; pp. 201–214.

31.   Duxson, P.; Fernández-Jiménez, A.; Provis, J.L.; Lukey, G.C.; Palomo, A.; Van Deventer, J.S.J. Geopolymer technology: The current state of the art. *J. Mater. Sci.* 2007, *42*, 2917–2933. https://doi.org/10.1007/s10853-006-0637-z

32.   Nergis, D.D.B.; Al Bakri Abdullah, M.M.; Sandu, A.V.; Vizureanu, P. XRD and TG-DTA study of new alkali activated materials based on fly ash with sand and glass powder. *Materials (Basel).* 2020, *13*. https://doi.org/10.3390/ma13020343

33.   Harja, M.; Barbuta, M.; Rusu, L. Obtaining and Characterization of the Polymer Concrete with Fly Ash. *J. Appl. Sci.* 2009, *9*, 88–96. https://doi.org/10.3923/jas.2009.88.96

34.   Burduhos Nergis, D.D.; Abdullah, M.M.A.B.; Sandu, A.V.; Vizureanu, P. XRD and TG-DTA Study of New Alkali Activated Materials Based on Fly Ash with Sand and Glass Powder. *Materials (Basel).* 2020, *13*, 343. https://doi.org/10.3390/ma13020343

35.   Burduhos Nergis, D.D.; Vizureanu, P.; Corbu, O. Synthesis and characteristics of local fly ash based geopolymers mixed with natural aggregates. *Rev. Chim.* 2019, *70*.

8. Green Materials Tendencies for a Sustainable Future

Sustainable development represents a multidisciplinary concept that combines social, economic and ecological aspects with the main goal to construct a livable system. This concept can basically be supported through the development, use and improvement of green materials [1]. Green materials are considered those products that use as raw materials regenerable sources and emphasis on reducing the use of hazardous substances in the design, manufacture and application. Also, when compared to conventional products, green materials present higher or comparable properties, lower production price and lower environmental impact. Therefore, these materials contribute to a more competitive industrial products and processes using the advanced materials design and manufacturing concepts – the aimed materials are obtained by low carbon emissions through an energy-efficient technology. During the processing of these newly developed materials, the emission of technological CO_2 is supposed to be very low compared to the traditional ones. Moreover, green materials exhibit socio-ecological benefits provided by products with higher integration level of functionality, lighter to transport, dynamic applications and decreased energy consumption, lowering the anthropogenic impact [2].

Green materials can also be defined as those environmentally friendly materials that are based on green chemical reactions (which take place between recycled or renewable components) and that involve the application of the principles of reduction or elimination of hazardous substances in their development, production and use [3][4]. At its basic level, research into ecological materials aims to develop alternative methods, processes or materials to traditional ones that offer comparable functional characteristics, but a much smaller impact on the environment. Sustainable development focuses on the positive impact of these materials, with a focus on reducing the use of hazardous substances in their design, manufacture and use.

8.1. Tendencies in geopolymers technology

Globalization, nowadays, generates large amounts of waste that significantly affects storage areas and living creatures in the vicinity. At the same time, the civil engineering sector is experiencing an exponential development process, which results in an increase in demand for building materials and usable space. Therefore, the need to obtain new materials was felt worldwide. One solution that has been intensively studied in the last past year, especially in this sector, consist in the development of environmentally friendly materials through a mechanism called geopolymerization. This multiple stage reaction is based on the chemical reaction between a waste, rich in aluminum and silicon, and an alkaline solution. The material resulting from geopolymerization results has tetragonal structure of Al-O-Si and properties comparable to those of Portland cement-based concrete.

For a sustainable development, the entire world must develop infrastructure adapted to specific climatic conditions, engineering, industrial and civil constructions or means of transportation by using materials with excellent properties and increasing the share of use of alternative materials and waste. This research direction will contribute to the reduction of energy consumption due to research both in the case of construction materials and means of transport can improve the national energy balance and the costs of imports of fossil fuel resources. Also, it will automatically produce the reduction of environmental harmful substances emission both in the construction and infrastructure sector and in the transport area, which have direct effects on the improvement of living conditions and costs related to health insurance.

Geopolymers are a complex class of ecofriendly materials. From one point of view, geopolymerisation represents an advantageous method of recycling aluminosilicate sources, but from another point of view, this technique increases and encourage the use of alkaline solutions. The synthesis of geopolymers is limited to the mixing of a reactive aluminosilicate material (metakaolin, fly ash etc.) with a solution of sodium, potassium or lithium hydroxide or silicate. The result is a low ordered structure, semi-crystalline, with a matrix of aluminosilicate gel phase which embeds multiple partial dissolved raw-material particles and multiple pores that contain residual liquid (water).

Geopolymers can be classified by many criteria, such as: used alkaline activator, properties possess before and after the hardening, minerals from structure, base materials used or application domain. The classification depending of the base materials used is presented in the below figure (Figure 9.1). The most used base materials in the early stages of developing geopolymers was the metakaolin. Later on, the investigations expanded to another materials, like calcinated clays, industrially wastes (ashes, slags, glass wastes, red mud) and many others natural materials reach in alumina and silica (zeolites, $Al_2O_3-2SiO_2$ powder or minerals with high magnesium content). In present, any materials with high silica and alumina content can be used as base materials to manufacture geopolymers.

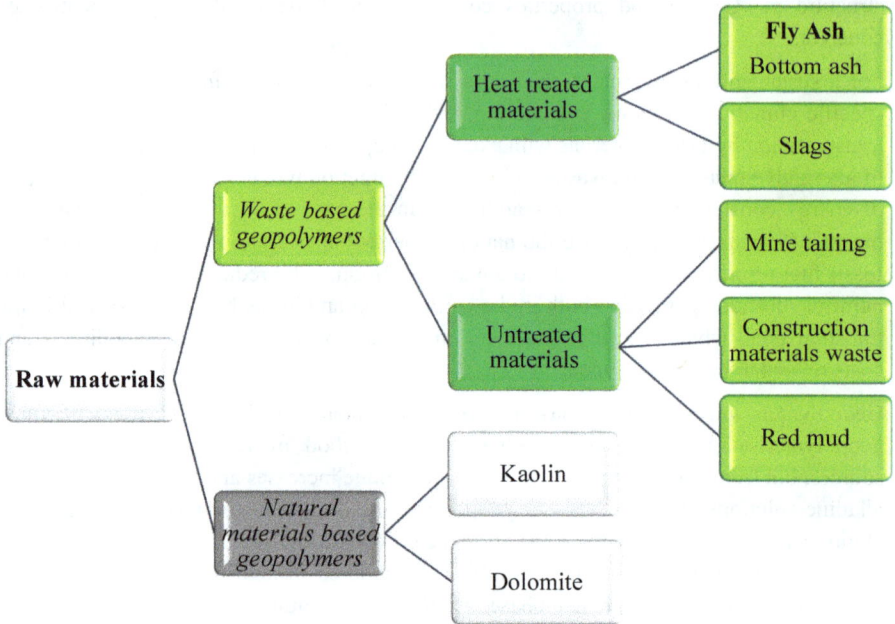

Figure 8.1. Geopolymers classification based on the aluminosilicate source.

The raw material plays a very important role in the development of the final characteristics of the hardened geopolymers at ambient temperature [5–7]. In particular, the presence of compounds or their concentration can lead to a sharp decrease in setting time (one of the most important parameters of the materials obtained at room temperature). The most important compound is calcium oxide, because, when mixed with sodium aluminate silicate, sodium silicate, but also calcium-alumino-sodium silicate, it can produce calcium silicate hydrate, the main component of materials based on Portland cement [8,9]. Therefore, a decrease in activation solution content or concentration could be obtained by increasing the CaO content in the geopolymers composition.

According to m. Gupta et al [10] the raw materials characteristics and the parameters of the activating solution greatly influence the mechanical properties of geopolymers. Therefore, in-situ cast of geopolymer composites, as replacement for Ordinary Portland cement materials, aren't feasible, especially, because the lack of standardizations.

Considering the characteristics of fly ash geopolymers, these types of composites are suitable for prefabricates manufacturing, such as bricks and panels. While the metakaolin based geopolymers develop better characteristics at lower temperature and lower activator consumption.

8.1.1. Geopolymers advantages

Both the properties and the quality of the geopolymers are strongly dependent on the type of base material, the source of this material and, of course, the quality of the main components. The main factors affecting the final properties of the geopolimer are: the source of Al-Si; activator type; the source and quality of the aggregate; water source; mixed quantities of each component; curing/drying time; curing temperature; particle size; calcium concentration; thermal treatment, if applied.

Geopolymers have a very wide range of features, compared to Portland Cement:

The first advantage is related to the source of the base material, any silicate source, pozzolanic or alumino-silicate compound that can be dissolved in an alkaline solution can be used as a source of geopolymer production.

Another important feature is low CO_2 emissions during manufacture. Geopolymers do not require high energy consumption (for the same amount of Portland cement, the ratio of energy used is 3/5 in favor of geopolymers).

Compared to conventional materials, some types of geopolymers are easy to prepare, being obtained only by mixing the rich material in aluminum and silicon with a strong alkaline solution, then cured at room temperature. A reasonable compressive strength is acquired shortly.

Compared to their properties, they have very good volumetric stability, low contraction (4/5 of that of Portland cement), high strength in a short time, they reach 70% of the ultimate compressive strength in the first 4 hours of curing. Geopolymers also have excellent durability, their properties are the same after decades, low thermal conductivity and extraordinary fire resistance, geopolymers retain their capabilities at temperatures up to 1000 °C - 1200 °C. The thermal conductivity of the geopolymers varies between 0.24 w/m*k and 0.3 w/m*k, compared with light refractory bricks (0.33 w/m*k up to 0.438 w/m*k). [21]

8.1.2. Geopolymers disadvantages

Some types of geopolymers can be considered "hard to create" because their manufacture requires special handling conditions of the components as well as one or more corrosive chemicals such as sodium hydroxide that can injure humans. Moreover, the particle size is very important, so the solid (aluminum-silicon) must be of superior quality.

Another disadvantage of geopolymers is that they can be sold only as pre-cast or pre-mixed materials due to the danger associated with their creation.

Pozzolanic geopolymeric cements possess high static and dynamic viscosity compared to Portland cement, so greater effort will be needed to remove the air voids from the cement paste.

Further research directions in the geopolymers field

The future research direction should be focused on the following lacks in geopolymers field:

- research on obtaining high-performance construction materials that lead to the reduction of infrastructure and construction maintenance costs;
- research on innovative technologies with low energy consumption for the manufacture of construction materials;
- research on the replacement of traditional raw materials used to obtain construction materials with waste or new materials with superior properties to reduce energy consumption and pollutants emitted in manufacturing processes in the commissioning and operation;
- -research regarding the obtaining of materials with special properties for the improvement of the structural and functional properties of the means of transport and of the afferent infrastructure for the reduction of fuel consumption and pollutants;
- developing of standardized testing methods and indicators in accordance with their composition and application;
- identification/optimization of activators with lower environmental impact than sodium silicate or hydroxide; alkaline solutions are definitely the highest disadvantage of geopolymers, especially, because they can produce harm to humans (personal protective equipment's are mandatory) and negatively affect the environment during production;
- optimization of mix design and parameters depending on the raw materials (aluminosilicate source);
- long term stability analysis and test (phase and structure stability over time); this type of evaluations could boost the introduction of geopolymers in industrial applications, especially, due to the fact that predictability is closely related with safety use;

- new area of application identification in accordance with their characteristics and tailored properties; knowing that geopolymers are oxidic, ceramic-like, materials with characteristics comparable with those of Ordinary Portland cement materials and tailored properties, when they are produced accordingly, they could be involved in cutting-edge applications;

- developing of geopolymers with new metallic or composite reinforcing elements; as for any composite material, the final characteristics are both influenced by the matrix properties and reinforcing elements properties, in the case of geopolymers these affirmation stands, therefore, in future geopolymers designs it must be considered the introduction of newly developed reinforcing elements;

- development of structures with high refractoriness and low thermal conductivity; even there are multiple studies regarding the obtaining of highly porous structure with low thermal conductivity, the final products aren't thermally stable or use additives in the production process;

- development of a semi-automatic technology (installations) for the production of geopolymeric products; up to our knowledge, in the present, there are no industrial installation that can be used for mass production of geopolymers, therefore, in the present, the researcher in the geopolymers field should focus their efforts on mass production and delivery from raw material to final destination (customer).

8.2. Tendencies in ferrock technology

Ferrock is a new green material, developed by David Stone [11], which can contribute to the reduction of CO_2 from the atmosphere. Ferrock is a cementitious material created through the chemical reaction between the CO and ferrous oxides which results in rock-like material based on iron carbonates. Basically, this material is five more times stronger to compressive strength than Portland cement and much more flexible and crack resistant, therefore it can successfully replace Portland cement in all its applications [12].

In present, this ecofriendly material is obtained from a blend of various materials, such as waste steel dust mixed with particles with high-content of silica oxides or mixes of iron powder with fly ash, limestone, metakaolin and oxalic acid. The chemical reactions on which the formation of this material is based are (eq. 8.1), (eq. 8.2) [13]:

$$Fe + 2CO^2 + H^2O \rightarrow Fe^{2+} + 2HCO^-_3 + H^2 \uparrow \tag{8.1}$$

$$Fe^{2+} + 2HCO_3^- \rightarrow FeCO_3 + CO_2 + H_2O \tag{8.2}$$

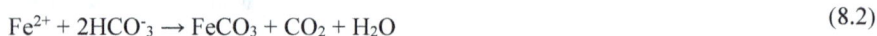

Therefore, the net reaction becomes (eq. 8.3):

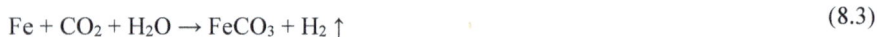

$$Fe + CO_2 + H_2O \rightarrow FeCO_3 + H_2 \uparrow \tag{8.3}$$

In their study, Mouli Prashanth P. et al [14], analysed the effect of molar concentration of oxalic acid on the compressive strength and carbonation depth of ferrock. According to their study, a ferrock made of 60 % iron powder, 20 % fly ash, 10 % limestone, 8 % metakaolin and 2 % oxalic acid (with molar concentration of 4, 6, 8, 10, respectively 12) exhibit a compressive strength value between 30 and 40 MPa at 7 days and between 40 and 60 MPa at 28 days. Also, it was observed that the increase of oxalic acid concentration up to 10 M produces an increase in compressive strength of ferrock, yet, above this concentration, i.e. at 12 M, the compressive strength is lower than at 10 M. The carbonation depth test performed on 28 days age samples showed that the ferrock samples have been fully carbonated compared to the control sample made of Ordinary Portland cement, concluding that ferrock blocks (Figure 8.2) absorbs considerable amount of CO_2 from the atmosphere.

a)   b)

Figure 8.2. Cork physical appearance and structure: a) macrostructure [15];
b) microstructure [16].

The mechanism of ferrock formation and the physico-chemical and chemical phenomena that occur during the formation of these materials are known only as base concepts. The reactions that happen between the unstable compounds after contact with carbon dioxide, oxygen or water have not yet been reported in current publications. Moreover, when other components are involved, such as metakaolin or fly ash and water, the geopolymerization reaction could take place. Basically, the mechanical, physical and chemical characteristics of the final product are influenced by multiple parameters, such as the concentration of each component, the molar concentration of oxalic acid, the ratio between the oxalic acid and the solid components, hardening conditions, mixing parameters, humidity and particle size distribution of each component, the sample volume, the geometry of the sample etc.

D. S. Vijayan et al [13] analysed the environmental impact of Ferrock made of 60 % iron powder, 20 % fly ash, 10 % limestone, 8 % metakaolin, by substituting 4 %, 8 % or 12 % of cement from the control sample (Ordinary Portland cement concrete) taking into account the energy and water consumption and the carbon depollution potential. According to their results, the best composition, in terms of compressive strength, flexural strength and environmental impact was the material with 8 % ferrock in its composition. Therefore, by introducing ferrock in Ordinary Portland cement concrete, a greener material can be obtained with better mechanical characteristics.

Further research directions in the ferrock field:
- with the present obtaining technology, the manufacturing costs of ferrock are higher than those of Ordinary Portland cement materials; therefore, a cost reduction analysis is necessary;
- as for geopolymers, a long-term stability analysis and test (phase and structure stability over time) is necessary, especially because in the field of ferrock very few publications can be found;
- developing of ferrock with new compositions, especially, the increase of recycled components is necessary.

8.3. Tendencies in Sorel cement technology

Magnesium binder compositions such as MgO — $MgCl_2$ — H_2O, are usually used in the manufacture of thermo-sound-insulating products, respectively for refractory products. Those materials are distinguished, in particular, by the method in which magnesium oxide is obtained and the ratio between the magnesium oxide and the salts used for activation. The most widespread magnesium binder is Sorel cement (also known as magnesia cement or magnesium oxychloride), obtained by using active magnesium oxide (caustic magnesite) resulting from the decomposition of magnesium carbonate at low temperatures. Sorel cement is a non-hydraulic cement, produced for the first time in 1867 by the chemist produced by the chemist Stanislas Sorel [1].

Magnesium cement was introduced at industrial level at the end of 19th century and the beginning of the 20th century. This material was mainly used for the manufacture of xylolite floors (self-supporting floor made of wood sawdust and binder [17]), as well as for faience and small architectural forms. In xylolite floors manufacturing sorel cement was used as binder filled with sawdust. Due to their low coefficient of abrasion, low thermal conductivity, seamless and apparently warm, during 50' they have been widespread in the entire Europe. However, due to the fact that the obtaining of this products required qualified personal and special handling operations, in the following years a much cheaper, available in huge quantities, was introduced on market, namely Ordinary Portland cement. Therefore, in the second part of the 20 centuries, the production of sorel cement decreased considerable.

The high reduction of sorel cement production was, also, facilitated other limitations, such as the lack of knowledge and researchers in this field. However, close to 90' the interest for the production and research in the field of this materials was resurrected. This phenomenon occurs when the special characteristics of magnesium cement, such as electromagnetic radiation absorbance, anti-electrostatic properties and incombustibility, where discovered. Magnesium binder and materials based on it have high strength characteristics, compared to the values of natural materials. Moreover, unlike natural materials, magnesium cement has possessed high tensile and bending strengths (comparable with Ordinary Portland cement materials, their main competitor), which is associated with the characteristics of solidified magnesite, in which magnesium oxychlorides are crystallizes as fibers (Figure 8.3). Fibrous crystals not only increase the strength of cement, but also act as a hardening element.

Due to its the high density, adhesion to organic materials, low alkalinity and the presence of bischofite mineral, the organic fillings do not rot in sorel cement, therefore, this material has antibacterial characteristics along with resistance to mold and fungus formation. Furthermore, when magnesium binders are used in construction mixtures, a dense, porous material with high wear, oil, gasoline and water resistance is obtained.

Sorel cement is a finely dispersed powder, the active part of which is magnesium oxide. One of its features is the need to use a special mortar - an aqueous solution of magnesium salts. The characteristics of this binder depend to a large extent on the accuracy of the dose of the components and on the observance of its rules of use. Magnesium cement is a worthy alternative to traditional Portland cement in mixtures of dry buildings, plaster mortars, flooring in industrial spaces, the manufacture of thermal insulators, glass and magnesium tiles.

Figure 8.3. Sorel cement microstructure [18].

The raw materials consist of crushed dolomite and magnesite carbonate rocks which are subjected firstly to a calcination stage at 800 °C and secondly to a grinding stage in order to obtain a fine powder.

To obtain the mixture, an aqueous solution of magnesium chloride is used. Mixing such a binder with water leads to a slow hardening of the material and a reduced strength. Increasing the magnesium sulfide content improves the water resistance of the finished product, but it affects the mechanical characteristics. Magnesium sulfide is more expensive than magnesium chloride. In the production of magnesium cement, it is important to find the optimal ratio between the components in order to obtain the ideal equilibrium between the production price and the product structural characteristics.

Also, when active magnesium oxides are used as secondary product in conventional concrete, a decrease in setting time will be obtained, along with an increase in bending, compression and tensile strength, at early stages.

According to J. Weber et al [18] the sorel cement belongs to the class of acid-base binders, which is produced through the reaction of calcinated magnesite (MgO) with magnesium chloride, $MgCl_2$. The resulting salt is reported to be a magnesium oxychloride hydrate with the formula $3MgO \cdot MgCl_2 \cdot 11H_2O$, probably in mixture with magnesium hydroxide, $Mg(OH)_2$, precipitated in a colloidal form. In the case of sorel cement, the initial setting time is of about 40 minutes, while the final setting time occurs after 9 hours, in normal boundary conditions. Regarding the compressive strength, they reported a tensile strength between 5 and 20 MPa and a compressive strength ranging from 20 to 100 MPa.

The main industrial application of magnesite cement is the production of seamless monolithic floors, including in industrial and public spaces. Those products are dust-free, abrasion-resistant, fire-retardant and, most important, durable. Modern magnesium floors are resistant to moisture due to the impregnation of their surface with water-repellent polymers. Unlike Portland cement floors, magnesium floors are crack resistant, chipped and beautiful. Also, sorel cement is used to fill floors with mosaic - bright, original, with unique patterns. Pieces of granite, marble and quartz chips are used as fillers in these cases. A coloring pigment is added to the composition. The artificial stone obtained from this binder is well polished, therefore the material is in high demand for the manufacture of small architectural forms, window sills and interior decorative elements.

However, the use of magnesite cement is limited by its negative properties. First of all, it is a low water resistance. Therefore, the binder is not used for operation in places with high humidity. Due to this disadvantage, magnesium cement has long been considered irrelevant for use in construction. But the discovery of new raw material depots and the development of the market for polymeric additives have given a boost to the expansion of magnesite cement production. Those because, the protection against moisture is provided by the introduction of special additives or waterproofing treatments applied on existing structures.

Further research directions in the ferrock field:
- developing of new compositions suitable for cutting-edge applications;
- much deeper research in structural and long-term stability;
- finding new sources of raw materials, considering the fact that periclase (MgO) and magnesite (MgCO3) are very limited, so their manufacture into Sorel cement is expensive and limited to small size products;
- introducing sorel cement in civil engineering applications, as replacement for Ordinary Portland cement.

8.4. Tendencies in cork technology

Cork is a natural renewable material with unique honeycomb microstructure made of small polyhedron cells filled with an air-like gas. The world cork production is about 340,000 tons/year, of which over 90 % of it is made in Portugal, Spain and Italy [19]. The first harvest is done when the tree reaches a minimum diameter of 0.7 m and at a height of 1.2 m from the ground. Only at the third harvest can a bark be obtained suitable for cork extracting, therefore, the first bark good for cork extraction is obtained after 40 years since the tree was planted. This material is considered ecofriendly especially due to the fact that the bark is renewable, its harvesting doesn't produce any risk for the plant, moreover, a new layer good for harvesting will be produced in about ten years.

According to H. Pereira [20], the structure of cork is different depending on the studying section plane, in the radial direction of tree the cells are prismatic and arranged in a brick-layered type, while in tangential plane, the cells are hexagonal. From geometric point of view, the cork tissue is axisymmetric. The SEM microstructure (Figure 8.4), shows that the solid mass is present only in the cell walls which define the structure regularity.

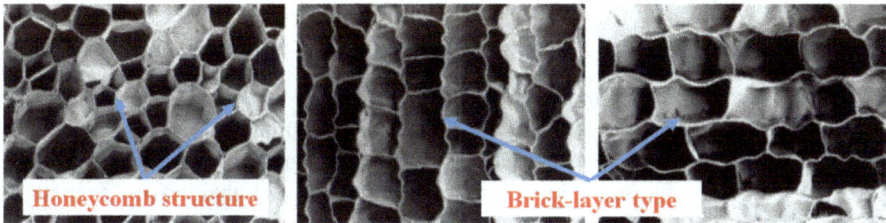

Figure 8.4. Cork microstructure in different sections [20].

Compared to other synthetic insulating materials, the thermal stability of cork is similar to that of polymers, especially due to the suberin and polysaccharides content. From compressive strength point of view, the cells absorb very much energy during densification, being fully densified after 85 % of deformation.

Cork is widely spread in multiple applications, from automotive production (chair covers), to furniture manufacturing, building insulating, boats manufacturing, inclined roof with rigid insulation over concrete slab, conical flat roof, traditional flat roof, green roof, inclined roof with roof membrane, inclined roof with corrugated roof systems, discontinuity between masonry and concrete walls, discontinuity between walls on metal structure and concrete, walls on metal structure with insulation, metal structure over masonry wall with insulation, double wall with insulation that fills the entire gap, interior walls insulated on both sides, partition layer for glass profiles, door leaf insulation, vibration control for heavy equipment, pipe and pipe insulation, expansion joints, formwork insulation, electric underfloor heating, traditional underfloor heating, filling cavities between beams floors, discontinuity between wall and screed, floating plate, floating plate wood, ventilated facades, double wall with insulation which partially fills the gap, exterior thermal and sound insulation of the walls. and wine bottling. Its application is related to its unique and special characteristics, such as lightweight, impermeable to liquids and gases, compressible, elastic, with very good thermal and acoustic insulating characteristics, high friction resistance etc. The properties of cork derive naturally from the structure and chemical composition of extremely strong and flexible cell membranes, which are waterproof and watertight. Because 89% of the bark tissue is composed of gaseous matter, the density of the cork is very low, air enclosed in the micro-cells is 90% by volume and about 50% by weight, the specific gravity ranges from 0.19 to 0.25 kg/m^3, which leads to a huge disproportion between the volume and the material weight. In 1 cm^3 of cork more than 40 million polyhedral with 14 sides are included, making cork five times lighter than water and unsinkable. The chemical composition of this materials includes: 45 % of suberin, 27% lignin, 12 % cellulose and polysaccharides, 6 % tannic acid, 5 % wax and other substances [21].

Further research directions in the cork field:
- developing of new binders with characteristics similar to those of cork suitable for high size products manufacturing; new area of applications can be developed by improving the thermal resistance and compressive strength of cork;
- evaluating the bonding characteristic and mechanics between the matrix and cork particles, when used as reinforcing elements.

8.5. Tendencies in sugarcane bagasse technology

Bagasse is the name given to the main waste (fibers) resulting from the extraction of molasses from sugar cane. These recyclable fibers consist of 40 - 60% cellulose, 20 - 30% hemicellulose and about 20% lignin [22]. Therefore, one of its applications (very widespread at the beginning) is as a fuel material for processing plants. However, with the growing interest in sustainability and materials recycling, especially natural ones, bagasse was introduced both in the civil engineering industry (in the manufacture of composite panels) and in the food industry (in the manufacture of cutlery or various types of packaging). This is mainly due to the elasticity and tensile strength of dry fibers, but also to the similarities with paper. Thus, with increasing restrictions on deforestation, the paper industry has begun to replace wood fiber with cane fiber, especially in the manufacture of napkins or cardboard [23]. Also, by mixing sugarcane bagasse with glycerol and tapioca starch high performance composites can be obtain, while ceramic and refractory materials can be obtained by mixing it with Arabic gum [24].

In their study, S.N. Monteiro et al [25] evaluated the possibility to obtain low cost composites, with resin matrix and bagasse as reinforcing elements. According to the process flow presented, composite plates of 150 mm x 250 mm x 6.5 mm with homogeneous structure can be cured at room temperature. However, the flexural strength (close to 20 MPa) of the products was under the expectation, mostly due to the poor adhesion between the polymeric matrix and the reinforcing elements. This behaviour it may be due to multiple debris and gummy tissues which remain sticked to the recycled fiber and doesn't allow to the resin to come in direct contact with the surface of the fiber (Figure 8.5). Therefore, for higher mechanical performances, a surface treatment for debris removing is necessary.

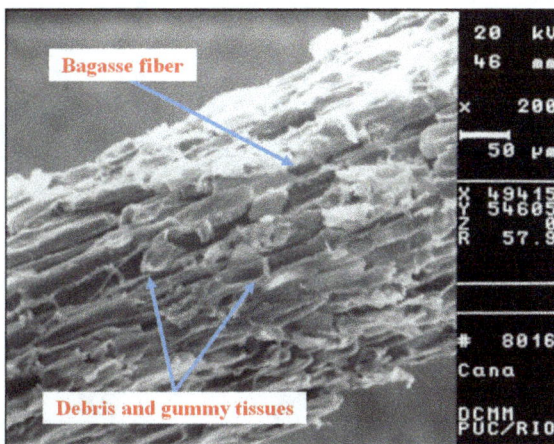

Figure 8.5. The microstructure of an untreated bagasse fiber [25].

In another study, D. Govindarajan et al [26] analysed the phasic and mineralogic composition of bagasse ash resulted from bagasse combustion in co-generation plant at an ethanol factory. The aim of this study was to evaluate the possibility to use this secondary product for partial substitution of cement in conventional concrete. According to their results, the use in concrete of the collected ash is limited by the high carbon content, however, after a suitable heat treatment, silica and calcium phases became primary, while the carbon content was significantly reduced.

Further research directions in the cork field:
- developing of new methods of fibers cleaning, in order to enhance the adhesion of matrix to bagasse fibers, when used as reinforcing elements;

- improving the burning technology of the sugarcane bagasse when used as combustion material, in order to decrease the carbon content of the resulted ash;

8.6. Tendencies in oriented strand board technology

A building material that has become an alternative to plywood or medium density fiberboard (MDF) is Oriented Strand Board wood (OSB). OSB boards are preferred in many civil engineering applications due to their mechanical characteristics, good insulating properties, low manufacturing costs and decorative appearance. OSB is a composite material made of wood chips, scrap from the manufacture of lumber, or even sawdust mixed with different types of adhesives (Figure 5).

Figure 8.6. The macrostructure of: a) OSB panel; b) MDF panel.

The main limitation for OSB application is related it its poor resistance to humidity and temperature. Those disadvantages are closely related with the type of chips (pine, bamboo, fir etc.) used along with the adhesive's (polymeric resins, wax etc.) adherence. Therefore, the actual research in this field is currently focused on the improvement of those two characteristics. In their study, M.N. Hornus et al [27] tested the impact on the stability of OSB panels of the pretreatment after hemicelluloses extraction. The treatment method consists of subjecting the pine chips to pressurized hot water in a Parr reactor. According to their study, the best wet conditions stability was obtained for the samples pretreated at 160 °C. However, considering the fact that this method is high cost energy consumption, greener methods are necessary.

One idea, that showed promising results was the adjustment of pressing stage parameters (temperature and force). X. Xiong [28] evaluated the effects produced by cold pressing and hot pressing on the characteristics of OSB. According to their results the samples obtained through the hot pressing method (110 °C for 4 minutes) showed better surface bonding strength (0.84 MPa), while the samples pressed at room temperature for 1 hour exhibit a value with 25% lower (0.63) MPa. Therefore, the hot pressing method produces better products in shorter time.

The board density, adhesive type and strand size influence on the mechanical characteristics was evaluated by R. Mirski et al [29]. According to their study the panels made with higher strands showed an increase of 15 % in bending strength, while an increase of 25 % was reported in tensile strength. However, the samples manufactured with 4.9 %, wt. of adhesive (melamine-urea-formaldehyde resin mixed with polymeric diphenylmethane glue) didn't present fluctuations in density.

Further research directions in the OSB manufacturing field:

- developing of new types of binders suitable for manufacturing of OSB with low moisture absorption;
- developing of engineered OSB structure with significant resistance to sun light.

References

1. Albino, V.; Balice, A.; Dangelico, R.M. Environmental strategies and green product development: An overview on sustainability-driven companies. *Bus. Strateg. Environ.* 2009, *18*, 83–96. https://doi.org/10.1002/bse.638

2. Bajpai, R.; Choudhary, K.; Srivastava, A.; Sangwan, K.S.; Singh, M. Environmental impact assessment of fly ash and silica fume based geopolymer concrete. *J. Clean. Prod.* 2020, *254*, 120147. https://doi.org/10.1016/j.jclepro.2020.120147

3. MISICK v. UNITED KINGDOM 165 ILR 544. In *International Law Reports*; Lauterpacht, E., Greenwood, C., Lee, K., Eds.; Cambridge University Press; 2016, pp. 544–553.

4. Komnitsas, K.A. Potential of geopolymer technology towards green buildings and sustainable cities. *Procedia Eng.* 2011, *21*, 1023–1032. https://doi.org/10.1016/j.proeng.2011.11.2108

5. Jindal, B.B. Investigations on the properties of geopolymer mortar and concrete with mineral admixtures: A review. *Constr. Build. Mater.* 2019. https://doi.org/10.1016/j.conbuildmat.2019.08.025

6. Hanjitsuwan, S.; Phoo-ngernkham, T.; Li, L. yuan; Damrongwiriyanupap, N.;

Chindaprasirt, P. Strength development and durability of alkali-activated fly ash mortar with calcium carbide residue as additive. *Constr. Build. Mater.* 2018, *162*, 714–723. https://doi.org/10.1016/j.conbuildmat.2017.12.034

7.   Rovnaník, P. Effect of curing temperature on the development of hard structure of metakaolin-based geopolymer. *Constr. Build. Mater.* 2010, *24*, 1176–1183. https://doi.org/10.1016/j.conbuildmat.2009.12.023

8.   McDonald, P.J.; Rodin, V.; Valori, A. Characterisation of intra- and inter-C-S-H gel pore water in white cement based on an analysis of NMR signal amplitudes as a function of water content. *Cem. Concr. Res.* 2010, *40*, 1656–1663. https://doi.org/10.1016/j.cemconres.2010.08.003

9.   Ng, C.; Alengaram, U.J.; Wong, L.S.; Mo, K.H.; Jumaat, M.Z.; Ramesh, S. A review on microstructural study and compressive strength of geopolymer mortar, paste and concrete. *Constr. Build. Mater.* 2018, *186*, 550–576.

10.   Gupta, M.; Kulkarni, N.H. A Review on the Recent Development of Ambient Cured Geopolymer Composites. In; Springer, Cham, 2020; pp. 179–188.

11.   Ferrock | Iron Shell Available online: https://www.ironshellmaterials.com/ferrock (accessed on Sep 28, 2020).

12.   Gallery | Iron Shell Available online: https://www.ironshellmaterials.com/gallery (accessed on Sep 29, 2020).

13.   D.S., V.D.S.A.T.S.J. Evaluation of ferrock: A greener substitute to cement. *Mater. Today Proc.* 2019, *22*, 781–787.

14.   Mouli, P.; Gokul, S. Investigation on ferrock based mortar an environment friendly concrete. *ternational Res. J. Eng. Technol.* 2019, 467–469.

15.   Ferrock: A Stronger, Greener Alternative to Concrete? Available online: https://buildabroad.org/2016/09/27/ferrock/ (accessed on Sep 30, 2020).

16.   Das, S.; Kizilkanat, A.B.; Chowdhury, S.; Stone, D.; Neithalath, N. Temperature-induced phase and microstructural transformations in a synthesized iron carbonate (siderite) complex. *Mater. Des.* 2016, *92*, 189–199. https://doi.org/10.1016/j.matdes.2015.12.010

17.   What is xylolite flooring? Available online: https://one-trade.co.uk/what-is-xylolite-flooring/ (accessed on Oct 2, 2020).

18.   Weber, J.; Bayer, K.; Pintér, F. Nineteenth century "novel" building materials: Examples of various historic mortars under the microscope. *RILEM Bookseries* 2013, *7*,

89–103. https://doi.org/10.1007/978-94-007-4635-0_7

19.   Despre Pluta – Jacork Available online: http://www.jacork.ro/despre-pluta/ (accessed on Sep 30, 2020).

20.   Pereira, H. The rationale behind cork properties: A review of structure and chemistry. *BioResources* 2015, *10*, 1–23. https://doi.org/10.15376/biores.10.3.Pereira.

21.   Variability of the Chemical Composition of Cork | Pereira | BioResources Available online: https://ojs.cnr.ncsu.edu/index.php/BioRes/article/view/3704 (accessed on Sep 30, 2020).

22.   Loh, Y.R.; Sujan, D.; Rahman, M.E.; Das, C.A. Review Sugarcane bagasse - The future composite material: A literature review. *Resour. Conserv. Recycl.* 2013, *75*, 14–22.

23.   Oliveira, J.A.; Cunha, F.A.; Ruotolo, L.A.M. Synthesis of zeolite from sugarcane bagasse fly ash and its application as a low-cost adsorbent to remove heavy metals. *J. Clean. Prod.* 2019, *229*, 956–963. https://doi.org/10.1016/j.jclepro.2019.05.069

24.   Doherty, W.; Halley, P.; Edye, L.; Rogers, D.; Cardona, F.; Park, Y.; Woo, T. Studies on polymers and composites from lignin and fiber derived from sugar cane. *Polym. Adv. Technol.* 2007, *18*, 673–678. https://doi.org/10.1002/pat.879

25.   Monteiro, S.N.; Rodriquez, R.J.S.; De Souza, M. V.; D'Almeida, J.R.M. Sugar cane bagasse waste as reinforcement in low cost composites. *Adv. Perform. Mater.* 1998, *5*, 183–191. https://doi.org/10.1023/A:1008678314233

26.   Govindarajan, D.; Jayalakshmi, G. XRD, FTIR and Microstructure Studies of Calcined Sugarcane Bagasse Ash. *Adv. Appl. Sci. Res.* 2011, *2*, 544–549.

27.   Hornus, M.N.; Cheng, G.; Erramuspe, I.V.; Peresin, M.S.; Gallagher, T.; Operations, F.; Via, B. Oriented strand board with improved dimensional stability by extraction of hemicelluloses. *Wood Fiber Sci.* 2020, *52*, 257–265. https://doi.org/10.22382/wfs-2020-024

28.   Xiong, X.; Ma, Q.; Ren, J. The performance optimization of oriented strand board veneer technology. *Coatings* 2020, *10*, 511. https://doi.org/10.3390/COATINGS10060511

29.   Radosław, M.A.D.D.D. Influence of Strand Size, Board Density, and Adhesive Type on Characteristics of Oriented Strand Lumber Boards Manufactured from Pine Strands | Mirski | BioResources. *BioResources* 2019, *14*, 6686–6696.

9. Conservation Potential by Aluminosilicates Recycling

9.1. Recycling potential and raw materials conservation

The recovery and reuse of recyclable resources are means of resolving the contradiction between the requirements of the economic growth process and the restrictive nature of resources [1].

Recycling is a concept of the twentieth century and has emerged as one of the ways to limit waste and to use resources more efficiently. It has become increasingly clear that industrialization and sustained population growth have led to the consumption increasing of resources amounts [2].

Therefore, many countries have already addressed the issue of recovering and recirculating reusable resources and have moved towards a unified coordination of this activity. The actions for the regulation of the recovery activities, the ones for the establishment of the organizational forms of collection, as well as the research efforts for finding the most efficient ways of recovery and capitalization of the reusable materials have been intensified [3,4].

In the field of recycling, the following strategies are required:

- Prevention of waste formation;
- Waste recovery by optimizing collection and sorting systems;
- Final disposal of waste that has not been recovered.

Recycling is only useful for the environment if the recycling process is conducted in a logical manner. By turning waste into usable resources, recycling provides a way to manage solid waste by reducing pollution, conserving energy, creating jobs and developing more competitive manufacturing industries. Just like dumping waste in specially designed areas or burning it in incinerators, recycling also costs money. The company's interest in recycling implies a full awareness of the economic and environmental benefits and costs of recycling, compared to the unilateral consumption of resources and storage of used products in specially designed areas or their burning in incinerators. When all these factors are taken into account, the benefits of recycling become apparent [5,6].

The greatest environmental benefit of recycling is not related to the storage of waste, but to the conservation of energy and natural resources and the prevention of pollution by using, in the manufacturing process, the materials resulting from recycling and less the primary ones. The recovered materials have already been previously purified and processed, so that their use in the manufacturing process implies a cleaner activity and a lower energy consumption. Detailed analyzes have shown that these environmental benefits of recycling are much more effective than any other environmental protection action [7].

Much less energy is required to turn recycled materials into new products, compared to starting production with raw, raw materials. By recycling a ton of materials in a regular recycling program, at least $ 187 are saved from electricity, oil, gas and coal, even when we take into account the consumption due to the collection and transportation of materials [5,8].

Recycling costs are partially amortized by avoiding storage or incineration costs and by selling the resulting materials. Storage prices vary widely by area, and the market for recycled materials is booming. However, recycling creates new jobs and increases the competitiveness of the manufacturing industry and provides the manufacturing industry with cheaper resources, long-term economic benefits that translate into value for consumers who spend less on products and packaging. The effects of recycling on industrial development are significant.

Moreover, recycling considerably reduces the dumping of waste at landfills which not only pollutes the environment massively but also creates a desolate image of cities, destroying the health of those who live around them. Therefore, the pollutants usually released into the water and air by depositing waste in landfills, is considerably reduced. Also, greenhouse gas emissions in the atmosphere are considerable reduced, especially, by replacing the used natural raw material with secondary raw material (the Earth's natural resources are preserved), resulting from recycling [9].

By recycling we save the community's energy costs in the long run, if we only think about the fact that the energy saved by recycling a single bottle can power a light bulb for four hours. Globally a wide range of mineral resources that are found in considerable quantities (existing and potential) in the soil and subsoil of the country, such as solid minerals such as gold, silver, salt, iron, copper, zinc, lead, uranium, but also those of oil and natural gas.

A related activity of the exploitation of mineral resources and which has, at the same time, a significant impact on the environment is the storage of waste from the exploitation of deposits. It is known that deposits of useful mineral substances are associated in the earth's crust with other minerals, which have no economic value, but often have to be extracted with useful rolls, for technical reasons, stability, consequence of being interspersed by rocks useful, etc. These minerals of no economic value are called generic gangue (tailings), becoming, after separation from useful mineral substances, mining waste. Proper management of this mining waste is one of the major challenges of the mining industry, both in terms of the significant impact and the environmental risks generated. Proper management of this mining waste is one of the major challenges of the mining industry, both in terms of the high costs it requires for handling and storage, and in terms of the significant impact and environmental risks generated.

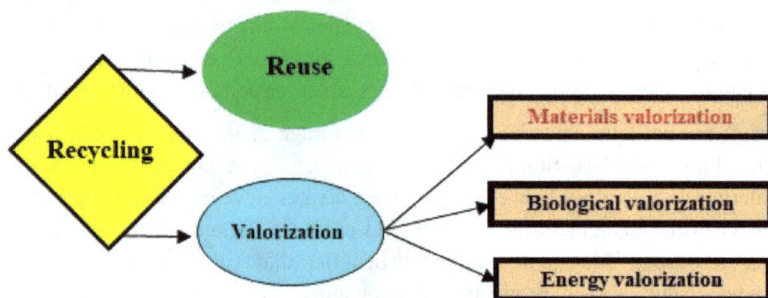

Figure 9.1. Recycling directions.

Better waste management can help: reduce greenhouse gas emissions (through reuse and recycling); increasing the efficiency of resource use - saving energy and reducing the consumption of materials through waste, prevention, reuse, recycling and recovery of energy from renewable sources; protecting public health through the safe management of potentially hazardous substances; protection of ecosystems (soils, groundwater, air emissions) [6,10,11].

Currently, the aim has been to promote the integrated management of waste resulting from the exploitation of ores and their recovery in the economy. The tailings dumps are mining waste that comes from the processing of useful minerals. Although these deposits are in fact minerals of no economic value if they are not included in a reuse / recycling process, they become very dangerous for the environment, especially in the event of natural hazards. Minimizing the risks on the environment, risks generated by the existence and improper storage of tailings, is considered an imperative that must be met.

Waste means technological waste, products and materials with expired warranty terms, physically used products that are no longer of use value, as well as household waste.

The waste of the manufacturing industry includes many different waste streams from a wide range of industrial processes, especially from: the production of base metals, food, beverages and tobacco products, wood and wood products, as well as paper products.

Industrial waste is on average 5 to 20 times larger than municipal waste and is stored in common pits with municipal waste or separately, there are other problems related to their inefficient management.

During 2010, the amount of waste generated by the extractive, energy and processing industry was 191 million tons, of which most (over 90%) waste from extraction activities (mining) - 240 million tons, and 15 million tons of waste generated by the energy and processing industry [12,13]. Most of the hazardous waste was disposed of by landfill, the rest being recovered or disposed of by co-incineration or incineration in the generators' own facilities or in specialized facilities belonging to private operators.

9.2. Replacing potential by mine tailings use

Useful minerals are the essential basis for any industrial activity and therefore the premise for the economic growth of any nation. Buildings, cars, ships, planes, glass, computers, are just a few examples of elements necessary for everyday life, the realization of which is impossible without the raw materials from mining. Globally, mining delivers more than 17 billion tons of raw materials annually, regardless of the building materials must meet the society's needs, and demand for raw materials will continue to grow in the coming decades due to the growth of the planet's population and the need of economic growth of developing countries [14]. The exploitation of useful mineral substances, whether those are carried out from up or underground, is one of the activities with an essential role in the development of modern society, a society in which almost half of the raw materials are mineral substances. The specificity of the mining activity is that it represents a temporary form of land use, these having other uses in the pre and post mining periods.

According to the information registered with the central authority for environmental protection, in 2002 344.5 million tons of waste were generated from mining activities. Of this total amount, most is waste from the excavation of coal, metalliferous and non-metalliferous ores, the so-called "mining tailings". This amount increased considerable over years, in 2018, the global tailing generation was estimated to total ~3.2 bn tons for copper and ~1.8 bn tons for iron [15].

Due to their specific characteristics, waste generated from mining activities is stored in deposits specially designed for this purpose, as follows:

- wastes from the exploitation and processing of metalliferous and non-metalliferous ores are deposited in tailings dumps and tailings ponds;
- waste from the exploitation of crude oil and natural gas is stored in batches.

Landfill is the most used method for disposing of industrial waste in Romania. Depending on the characteristics of the waste, storage is done in two main ways:

- non-hazardous industrial waste is usually disposed of, on a contract basis, at municipal landfills, which are managed by the local public administration;
- hazardous industrial waste or those that are not accepted on municipal landfills are disposed of in landfills of their own generating economic agents.

Depending on the nature of the waste deposited, industrial waste landfills have been classified as follows:

a) tailings dumps - land surfaces on which the material resulting from the excavation of non-metalliferous and metalliferous ores is deposited;

b) slag and ash dumps - land surfaces on which the material resulting from thermal processes is deposited (burning coal in thermal power plants, metallurgical processes);

c) battles - excavated land surfaces in which liquid waste is generally deposited; the term is used both for hazardous waste generated from the activities of oil refining and processing of petroleum products, and for sludge, waste from the chemical industry, waste generated from animal husbandry activities;

d) tailings ponds - excavated land surfaces in which liquid waste with a high content of suspensions is deposited, in order to sediment them; the term is used both for semi-liquid waste from mining activities and for liquid waste generated in the chemical industry, food industry etc.;

e) drying beds - represent the installations related to an industrial wastewater treatment plant where the sludge is deposited and where its natural dehydration takes place;

f) simple industrial landfills - represent the land surfaces arranged at the soil surface for solid waste disposal;

g) underground landfills - underground arrangement for waste storage.

In times when raw material resources were dwindling, landfills in ancient cities contained less waste with potential for recycling (tools, pottery, etc.).

In Europe, in pre-industrial eras, waste from the processing of bronze and other precious metals was collected and melted for continuous reuse, and in some areas dust and ash from coal or wood fires were reused to obtain the basic material in the manufacture of bricks. The main reason for practicing material recycling was the economic advantage, the need for natural raw materials thus becoming lower [16].

The mining industry exerts special influences on the environment, which is manifested in all phases of technological production processes.

Mining generates an impact that affects, to a greater or lesser extent, all environmental factors. Depending on a whole series of variables, the impact can be permanent or temporary, reversible or irreversible, negative or positive. On the other hand, the mining industry has at hand the technique and technologies necessary for the ecological rehabilitation of degraded lands and the elimination of impact, within the practice of responsible mining, focused on the three basic pillars of sustainable development: economic development, environmental protection, social protection. Adding to these pillars and technological progress, mining can become a sustainable activity in the medium and long term. Mining activities, as a whole, do not fall indefinitely in the context of sustainable development, but the succession of mining operations, namely exploration, preparation, exploitation, closure and ecological rehabilitation can be directed so that the environment, economy and local community reach standards of sustainability. clearly superior to those prior to the development of mining [17].

The severity of the problems related to the influence of the mining industry on the environmental factors requires both the designers and those who will lead the productive activities in the field, to anticipate the negative effects, and to take all possible measures of prevention, protection and recovery.

The tailings dumps contain a wide variety of rocks and soil depending on the geology and type of mine. Formally, waste deposited in tailings dumps can be classified as follows:

- sterile material and rocks removed from surface quarries;
- sterile material and rocks from underground mines, very poor ores that are not processed;
- process waste - dry waste from initial processing, such as large or small material, coarse material from wet processing, other residues.

The quantity and characteristics of the waste deposited in the tailing's dumps vary depending on the type of ore mined and the specific way of carrying out the activity.

The organization phase of the mining units requires the execution of specific activities (arrangement of access roads and connection with existing ones, construction of work platforms, construction of premises and sometimes modification of natural drainage), each of which is elements of disturbance, modification and interruption. of the continuity of the environment.

The extraction of useful mineral substances produces obvious effects when done with explosives, either by noise pollution (noise) or by emitting large amounts of dust, which causes major damage to vegetation in nearby areas. Extraction by mechanical means produces a noise pollution due to the operation of the machines (permanent noise). Other problems can be caused by the dredging of alluvial materials, when irreversible alterations of aquatic habitat can occur, in terms of physical, chemical, biological, with consequences both in the upstream and downstream areas.

The transport and processing of the extracted materials causes, above all, noise pollution and dust emanation, with effects on the vegetation and fauna of the area. Another activity responsible for various changes in the physico-chemical characteristics of water and river habitat is the discharge into the watercourses of sludge residues from preparation plants.

The discovery of a deposit, the extraction of useful mineral substances, transport and storage are destructive actions, with repercussions on the local habitat and fauna. These effects are extremely serious in the case of high-value natural environments, as they endanger ecosystems containing rare or protected species of flora and fauna. Other major issues are the need to relocate the resident population or the diversion of watercourses in the case of the exploitation of deposits of mineral substances useful in the quarry.

9.3. Replacing potential by red mud use

Red mud is a waste from the Bayer bauxite refining process. The impact on this environment is given by the alkalinity so the composition, pH, buffer capacity and its causes were followed. In the red mud there are other constituents that are bases according to Lewis's definition and that contribute to the increase of alkalinity. Red sludge can be stored in lagoons, in liquid or dry form. The lower the adherent liquid content, the lower the amount of NaOH. The aquatic environment can be affected by both types of storage by leachate [18,19].

The large area occupied by this waste, are mainly related with the deposition strategy, which consist, mainly, for the first half of the 20th century, in the disposal of the resulted residue from the Bayer process in land-based ponds [20,21]. Due to the fact that the market request for alumina increased constantly, the quantity of the deposited residues presented the same trend (Figure 9.2). Therefore, globally during '60, the total amount of red mud deposited in lakes reached about 200 million tons. Even, during the past few years, few smart recycling methods have been developed, the amount of waste which is being deposited annually is alarming [22].

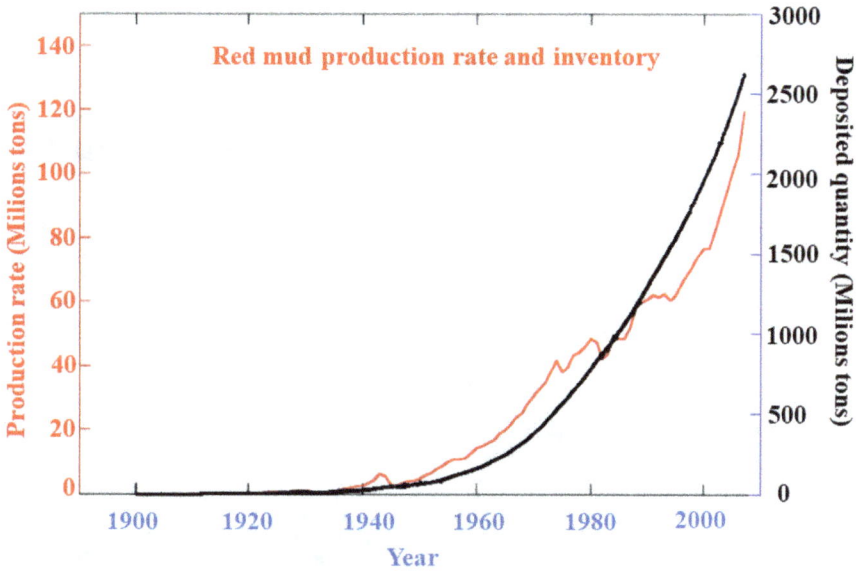

Figure 9.2. Red mud global production rate and deposited quantity [23].

The chemical composition of red mud deposits varies depending on a multitude of factors, but especially depending on the composition of the raw material (bauxite). However, in essence the red mud is a strong alkaline waste due to the complete non-removal of NaOH, in the storage sites it is impossible to restore the fauna and flora [24]. Therefore, the only way to recover the space occupied by it is to use this waste in the manufacture of useful materials.

The influence on environmental factors begins with the activity of prospecting and exploration of deposits, and continues and intensifies with the development of productive activities. In some cases, the negative influence manifests itself for a very long time, even after the total cessation of productive activity in the area.

9.4. Replacing potential by fly-ash use

Fly ash is a secondary product of coal burning in thermal power plants, this waste is considered a highly environment polluter due to its spreading by wind on the land in the close areas of power plants and waste-dumps areas. Since the need for energy is increasing, until the clean energy production will reach 100% of the energy request, the power plants will produce more coal ash. In 2008 the worldwide total amount of coal ash produced was estimated to approximately 900 million tons. Moreover, it is expected that the total amount of coal ash will increase substantially reaching 2000 million tons by 2020. Therefore, it is highly necessary to create and develop alternative technology that this waste as raw material in order to reduce the negative environmental impact [25,26].

Thermal power plant ash is unanimously classified as industrial waste with severe ecological impact, most of it is resulted from electricity production flows which are based on the burning of coal. Globally, its production takes place massively, in small geographical area, which are strongly related with the areas were coal reserves are discovered.

Fly ash production (million tons/year)

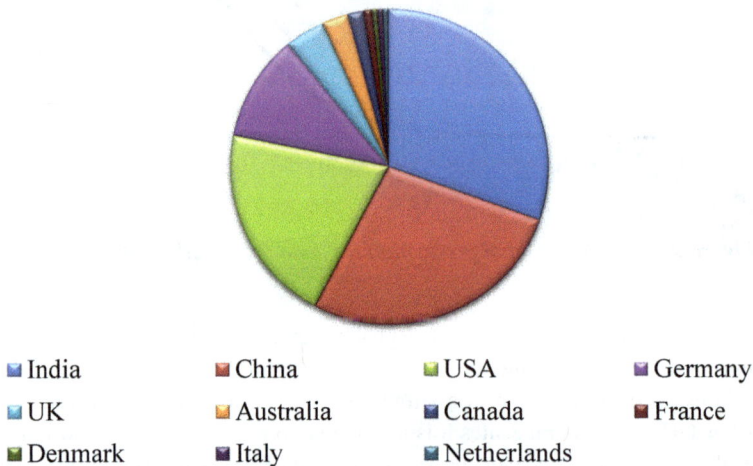

⌷ India	⬛ China	⌷ USA	⬛ Germany
⌷ UK	⌷ Australia	⬛ Canada	⬛ France
⬛ Denmark	⬛ Italy	⬛ Netherlands	

Figure 9.3. Fly ash production in different countries [27].

Due to the fact that fly ash is a silico-aluminium or low calcium material which can be converted by geopolymerisation in high performance materials suitable for many applications, especially in construction industry as concrete, pavements, tank for containment and immobilization of radioactive wastes, but, also as refractory ceramic-like materials due to its elemental structure. Due to its great properties, such as chemical and fire resistance, very good mechanical strength, fly ash is a waste material that have a great potential in utilization as base material for many applications.

References

1. Cleveland, C.J.; Morris, C.G. *Handbook of Energy: Chronologies, Top Ten Lists, and Word Clouds*; Elsevier, 2013; ISBN 978-0-12-417019-3.

2. Goodman, P.S. Where Gadgets Go To Die: E-Waste Recycler Opens New Plant in Las Vegas. *Huffingt. Post* 2012.

3. Sahni, S.; Gutowski, T.G. Your scrap, my scrap! the flow of scrap materials through international trade. In Proceedings of the Proceedings of the 2011 IEEE International Symposium on Sustainable Systems and Technology, ISSST 2011; 2011; pp. 1–6.

4. Sortation by the numbers. *Resour. Recycl. News* 2018.

5. Vigsø, D. Deposits on single use containers - A social cost-benefit analysis of the Danish deposit system for single use drink containers. *Waste Manag. Res.* 2004, *22*, 477–487. https://doi.org/10.1177/0734242X04049252

6. *Puzzled About Recycling's Value? Look Beyond the Bin*; United States Environmental Protection Agency, 1998.

7. Zaman, A.U.; Lehmann, S. Challenges and Opportunities in Transforming a City into a "Zero Waste City." *Challenges* 2011, *2*, 73–93. https://doi.org/10.3390/challe2040073

8. The truth about recycling. *Econ.* 2007.

9. Clark, B.; Foster, J.B. Ecological imperialism and the global metabolic rift: Unequal exchange and the guano/nitrates trade. *Int. J. Comp. Sociol.* 2009, *50*, 311–334. https://doi.org/10.1177/0020715209105144

10. Brown, M.T.; Buranakarn, V. Emergy indices and ratios for sustainable material cycles and recycle options. *Resour. Conserv. Recycl.* 2003, *38*, 1–22. https://doi.org/10.1016/S0921-3449(02)00093-9

11. Pimenteira, C.A.P.; Pereira, A.S.; Oliveira, L.B.; Rosa, L.P.; Reis, M.M.; Henriques, R.M. Energy conservation and CO 2 emission reductions due to recycling in Brazil. *Waste Manag.* 2004, *24*, 889–897. https://doi.org/10.1016/j.wasman.2004.07.001

12. Waste generated and treated in Europe, Available online: https://ec.europa.eu/eurostat/documents/3217494/5646109/KS-55-03-471-DE.PDF/744c49e5-e7ca-4c3b-bae3-611c5ccb43c2?version=1.0 (accessed on Oct 7, 2020).

13. Commission, E. *EU Waste Legislation*; 2014;

14. Moukannaa, S.; Loutou, M.; Benzaazoua, M.; Vitola, L.; Alami, J.; Hakkou, R. Recycling of phosphate mine tailings for the production of geopolymers. *J. Clean. Prod.*

2018, *185*, 891–903. https://doi.org/10.1016/j.jclepro.2018.03.094

15.   How to generate value from tailings with reprocessing to be future ready? - Metso Available online: https://www.metso.com/blog/mining/how-to-generate-value-from-tailings-with-reprocessing-to-be-future-ready/ (accessed on Oct 7, 2020).

16.   Council, E. *The Producer Responsibility Principle of the WEEE Directive*; Available online: https://ec.europa.eu/environment/waste/weee/pdf/final_rep_okopol.pdf (accessed on Oct 7, 2020).

17.   Ayres, R.U.; Holmberg, J.; Andersson, B. Materials and the global environment: Waste mining in the 21st century. *MRS Bull.* 2001, *26*, 477–480. https://doi.org/10.1557/mrs2001.119

18.   Pepper, R.A.; Couperthwaite, S.J.; Millar, G.J. Comprehensive examination of acid leaching behaviour of mineral phases from red mud: Recovery of Fe, Al, Ti, and Si. *Miner. Eng.* 2016, *99*, 8–18. https://doi.org/10.1016/j.mineng.2016.09.012

19.   Wang, L.; Sun, N.; Tang, H.; Sun, W. A review on comprehensive utilization of red mud and prospect analysis. *Minerals* 2019, *9*.

20.   Power, G.; Gräfe, M.; Klauber, C. Bauxite residue issues: I. Current management, disposal and storage practices. *Hydrometallurgy* 2011, *108*, 33–45. https://doi.org/10.1016/j.hydromet.2011.02.006

21.   *Bauxite Residue Management: Best Practice*; World Aluminum; available online: http://bauxite.world-aluminium.org/fileadmin/user_upload/Bauxite_   Residue_ Management _-_Best_Practice__English__Compressed.pdf (accessed on Oct 7, 2020).

22.   Enserink, M. After red mud flood, scientists try to halt wave of fear and rumors. *Science (80-.).* 2010, *330*, 432–433.

23.   Craig, K.; Markus, G.; Greg, P. Review of Bauxite Residue "Re-use" Options. *Proj. ATF-06-3 "Management Bauxite Residues"* 2009.

24.   Mayes, W.M.; Jarvis, A.P.; Burke, I.T.; Walton, M.; Feigl, V.; Klebercz, O.; Gruiz, K. Dispersal and Attenuation of Trace Contaminants Downstream of the Ajka Bauxite Residue (Red Mud) Depository Failure, Hungary. *Environ. Sci. Technol.* 2011, *45*, 5147–5155. https://doi.org/10.1021/es200850y

25.   Singh, D.N.; Kolay, P.K. Simulation of ash-water interaction and its influence on ash characteristics. *Prog. Energy Combust. Sci.* 2002, *28*, 267–299.

26.   Ahmaruzzaman, M. A review on the utilization of fly ash. *Prog. Energy Combust. Sci.* 2010, *36*, 327–363.

27.   Aakash, D.; Manish Kumar, J. Fly ash – waste management and overview: A Review. *Recent Res. Sci. Technol.* 2014, *6*, 30–35.

About the Authors

Petrica VIZUREANU
Professor Ph.D. Eng.
Head of department at Department of Technology and Equipment for Materials Processing
Faculty of Materials Science and Engineering, "Gheorghe Asachi" Technical University of Iasi
peviz2002@yahoo.com

Professor and researcher at "Gheorghe Asachi" Technical University of Iasi, with more than 30 years of experience. Ph.D. degree, since 1999 in Materials science and engineering; 2010 - present Ph.D. Supervisor in Materials Engineering domain. He has over 150 publications, 130 articles being indexed in ISI Web of Science. He has large experience in the field of composite materials; ceramic materials, insulating materials; optimization of materials characteristics. H-index is 17.

Dumitru-Doru BURDUHOS-NERGIS
Assistant Professor PhD. Eng.
Department of Technology and Equipment for Materials Processing
Gheorghe Asachi Technical University of Iasi
Bunduc.doru@yahoo.com / doru.burduhos@tuiasi.ro

Researcher and assistant professor at Gheorghe Asachi Technical University of Iasi, Faculty of Materials Science and Engineering. Graduates the Advanced techniques of materials processing engineering master programme with the thesis entitled „Studies on obtaining new aluminosilicate materials" and continues the studies in the composites materials field during the Ph.D. on thesis entitled "Contributions in obtaining geopolymers from mineral wastes" under the scientific coordination of Prof. Ph.D. Eng. Petrică VIZUREANU. Since 2016 he has multiple research papers published in conferences volumes or in WOS indexed Journals. H-index is 6.